WITHDRAWN

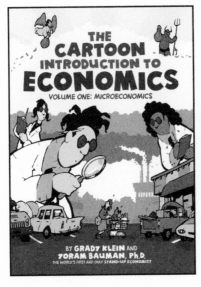

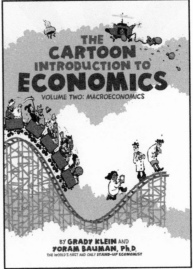

THE
CARTOON
INTRODUCTION TO
STATISTICS

THE CARTOON INTRODUCTION TO STATISTICS

BY **GRADY KLEIN** AND **ALAN DABNEY, Ph.D.**

A NOVEL GRAPHIC FROM HILL AND WANG
A DIVISION OF FARRAR, STRAUS AND GIROUX
NEW YORK

HILL AND WANG
A DIVISION OF FARRAR, STRAUS AND GIROUX
18 WEST 18TH STREET, NEW YORK 10011

PRINTED IN THE UNITED STATES OF AMERICA
PUBLISHED SIMULTANEOUSLY IN HARDCOVER AND PAPERBACK
FIRST EDITION, 2013

LIBRARY OF CONGRESS CATALOGING-IN-PUBLICATION DATA
KLEIN, GRADY.
    THE CARTOON INTRODUCTION TO STATISTICS / BY GRADY KLEIN AND
ALAN DABNEY, PH.D. — FIRST EDITION.
        P. CM.
    ISBN 978-0-8090-3366-9 (HARDCOVER) — ISBN 978-0-8090-3359-1 (TRADE PBK.:
ALK. PAPER)
    1. MATHEMATICAL STATISTICS — COMIC BOOKS, STRIPS, ETC.   2. GRAPHIC NOVELS.
    I. DABNEY, ALAN, 1976 –   II. TITLE.

QA276 .K544 2013
519.5 — DC23
                                                      2012030027

WWW.FSGBOOKS.COM
WWW.TWITTER.COM/FSGBOOKS • WWW.FACEBOOK.COM/FSGBOOKS

1 3 5 7 9 10 8 6 4 2

CONTENTS

MOST OF US ENCOUNTER STATISTICS **EVERY DAY**...

AWESOME!

ONE BOWL OF CHOCOLATY FROSTO BOMBS HAS 1,200% OF MY DAILY RECOMMENDED SUGAR.

...EVEN IF WE DON'T **CRUNCH NUMBERS** FOR A LIVING.

STATISTICS RADIATE FROM OUR **TELEVISIONS**...

THIS SHOW HAS AN **ESTIMATED AUDIENCE OF 4.8 MILLION!**

IT **MUST** BE GOOD.

...EMANATE FROM OUR **PHONES**...

YOU SENT MORE **TEXT MESSAGES** THIS MONTH THAN THE **ENTIRE NATION OF CHAD.**

...SEEP FROM OUR **RADIOS**...

POLLS SHOW SENATOR NEERDORPH WITH A **40-POINT** LEAD!

...AND LITTER OUR **ROADS**.

KILLED OR **INJURED?**
Average Award $150 million, and Beyond!

Eaton Cake, Attorney at Law
800 WIN CAKE

SCOURADE!

78% of Dentists Recommend It!

IT'S **IMPOSSIBLE TO ESCAPE THEM.**

THEY'RE **EVERYWHERE**!

AT THE **MALL**

WE'RE PLAYING THAT BACKGROUND MUSIC...

...BECAUSE STUDIES SHOW IT'LL MAKE YOU BUY **10% MORE CLOTHES**!

AT **SCHOOL**

YUP, I'M **GRADING ON A CURVE**!

IN THE **KITCHEN**

WHY DO I HAVE TO DO THE DISHES, LIKE, 75% OF THE TIME?

BECAUSE I COOK DINNER 99% OF THE TIME.

IN THE **BEDROOM**

THIS WEBSITE WILL FIND MY **PERFECT MATCH**...

...IF I ENTER MY VITAL STATISTICS.

STATISTICS ARE WITH US **WHEN WE'RE BORN**...

95% OF ALL BABIES ARE DELIVERED BETWEEN WEEKS 38 AND 42...

...SO THAT'S WHEN WE'LL DELIVER YOURS.

...AND LIKE IT OR NOT, WE WILL **BECOME STATISTICS** WHEN WE DIE.

SO SAD.

AT LEAST HE LIVED **LONGER THAN THE AVERAGE BEAGLE**.

FORTUNATELY, THERE'S A **GOOD REASON** FOR ALL THIS.

STATISTICS ARE EVERYWHERE **BECAUSE THEY'RE SO USEFUL.**

STATISTICS HELP PEOPLE PREDICT THE **WEATHER...**

THERE'S A **95%** CHANCE IT'LL BE SUNNY TOMORROW!

BUT ALSO A **3%** CHANCE IT'LL RAIN FROGS!

...AND ORGANIZE THE **INTERNET...**

BASED ON YOUR BUYING HISTORY, I'VE GOT **RECOMMENDATIONS** FOR YOU.

HOW DID IT KNOW I WANTED A WILLIAM SHATNER DOLL?

 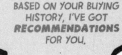 

...AND DEVELOP **MEDICINES...**

OUR STUDIES SHOW THAT THIS PILL IS ONLY **2.5%** MORE LIKELY THAN A PLACEBO TO PREVENT CANCER, WITH A MARGIN OF ERROR OF **12%**...

...BUT IT WORKS GREAT AS A **LAXATIVE!**

GREAT, WHAT SHOULD WE CALL IT?

...AND **INFLUENCE FASHION.**

LIKE, WOW YOU ARE SO **1987,** I LOVE IT!

BUT LOSE THE BELL-BOTTOMS.

I USED STATISTICS TO DETERMINE THAT **DENIM JACKETS** ARE LIKELY TO COME BACK IN STYLE THIS YEAR...

AND THAT'S **NOT ALL.**

SO **WHAT MAKES** STATISTICS **SO INCREDIBLY USEFUL?**

THIS THING IS **AMAZING!**

IT'S GOT A **FORK** AND A **KNIFE** AND A **SPOON** AND A **RAKE** AND A **STRAW...**

...AND A **DRILL** AND **FINGERNAIL CLIPPERS** AND A **PENCIL** AND...

THE **SIMPLE ANSWER** IS THAT STATISTICS HELP US COME TO GRIPS WITH **LARGE NUMBERS OF IMPORTANT THINGS...**

94% OF ALL HUMANS WHO EVER LIVED **ARE DEAD...**

...AND **200** MILLION OF THEM **DIED IN THE PLAGUE...**

...AND TRAFFIC **KILLS MILLIONS MORE** EVERY YEAR,...

...AND YOUR ODDS OF BEING **STRUCK BY LIGHTNING** ARE WAY **HIGHER** IF YOU PLAY GOLF!

...WHICH IN TURN CAN HELP US **BETTER UNDERSTAND OUR COMPLEX WORLD...**

...AND **MANIPULATE IT.**

STUDIES SUGGEST THAT 78% OF ALL PEOPLE **LIKE** DOUGHNUTS.

SO IF WE GIVE THEM AWAY **FREE** AT OUR DEATH CULT MEETINGS...

...WE CAN **ATTRACT MORE MEMBERS!**

BUT THE **REAL POWER** OF STATISTICS IS **MORE SPECIFIC.**

HERE'S THE **REAL REASON** EVERYONE USES STATISTICS:

STATISTICS HELP US **MAKE CONFIDENT DECISIONS...**

...WHEN WE HAVE **LIMITED INFORMATION.**

BUT LET'S EXPLAIN WHAT THAT MEANS...

IMAGINE THAT WE WANT TO KNOW THE **AVERAGE WEIGHT**...

...OF **ALL THE FISH IN A LAKE**.

HERE FISHY, HERE FISHY, FISHY...

IF WE FIND OUT HOW MUCH EACH FISH WEIGHS ON AVERAGE...

...WE CAN FIGURE OUT ABOUT HOW MANY FISH WE NEED TO CATCH EACH DAY TO **KEEP OUR CATS FROM STARVING!**

IF WE **DRAINED THE LAKE** AND **WEIGHED EVERY SINGLE FISH**...

...WE'D HAVE **ALL THE INFORMATION** WE NEED TO CALCULATE THE ANSWER.

BUT FOR OBVIOUS REASONS, WE **CAN'T DO THAT**.

MAYBE THAT WASN'T SUCH A **GOOD IDEA**.

ON THE OTHER HAND, IF WE CATCH A **SAMPLE** OF **100 FISH** AND WEIGH THEM...

THESE **100** FISH WEIGH **247** POUNDS.

SO THE **AVERAGE FISH IN THIS SAMPLE** WEIGHS **2.47** POUNDS!

... WE'LL HAVE ONLY **LIMITED INFORMATION** ABOUT ALL THE FISH.

SO NOW WE KNOW THE AVERAGE WEIGHT **IN THIS** SAMPLE PILE...

...BUT WE STILL DON'T KNOW THE AVERAGE WEIGHT OF THE **REST OF THE FISH IN THE LAKE.**

BUT HERE'S THE **COOL THING:**

WITH THE TOOLS OF STATISTICS WE CAN **USE THIS LIMITED INFORMATION...**

... TO MAKE **CONFIDENT STATEMENTS** ABOUT **ALL** THE FISH IN THE LAKE.

STATISTICS IS ABOUT **USING THE FISH WE DID CATCH...**

... TO **SAY THINGS ABOUT THE FISH WE DIDN'T.**

REALLY? HOW DOES **THAT WORK?**

THAT'S WHAT THIS BOOK IS ABOUT!

THIS BOOK IS ABOUT THE **FUNDAMENTAL QUESTION** OF STATISTICS:

HOW DO WE USE **SAMPLES...**

ALONG THE WAY WE'LL LEARN TO SIFT THROUGH **BIG PILES OF DATA**...

... CALCULATE **CONFIDENCE INTERVALS**...

... AND **TEST HYPOTHESES.**

ARGH!

WE'RE **SKEWED!**

THAT'S **NOT NORMAL!**

I'M 95% CONFIDENT THAT WE HATE YOU ABOUT THIS MUCH.

I'M 3% CONFIDENT THAT MY EVIL MACHINE **ISN'T BUSTED!**

AND MORE GENERALLY, WE'LL GET A SENSE OF THE KIND OF THINGS YOU **CAN**...

... AND **CAN'T**...

... DO WITH STATISTICS.

WE CAN USE STATISTICS TO **MAKE CONFIDENT GUESSES**...

... BUT YOU CAN **NEVER** USE THEM TO **ACHIEVE CERTAINTY.**

IF WE DON'T CATCH **ALL** THE FISH...

... WE'LL NEVER KNOW **FOR CERTAIN** WHAT'S DOWN THERE.

13

IN THIS BOOK WE'RE GOING TO **FOCUS ON THE BASIC CONCEPTS.**

LIKE **STANDARD DEVIATIONS...**

...AND **SAMPLING DISTRIBUTIONS...**

...AND **PROBABILITIES...**

...AND **CONFIDENCE!**

BUT IF YOU'RE ALSO CURIOUS ABOUT THE **TECHNICAL DETAILS...**

...YOU CAN FIND THOSE IN A SECTION AT THE END CALLED **THE MATH CAVE.**

LIKE WHAT THE HECK DO THESE **FORMULAS** AND **SYMBOLS** MEAN?

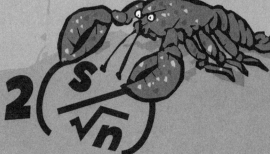

$$\bar{x} \pm 2\left(\frac{s}{\sqrt{n}}\right)$$

PART ONE
GATHERING STATISTICS

NO PEEKING.

CHAPTER 1
NUMBERS

IN THIS CORNER, WEIGHING IN AT
50.8 TRILLION NANOGRAMS...

...THE **DWARF!**

AND IN THIS CORNER,
WEIGHING IN AT
0.193 TONS...

...THE **GIANT!**

AS WE LEARNED IN THE INTRO, STATISTICS ISN'T **JUST ABOUT NUMBERS.**

GOOD MORNING, ACCOUNT NUMBER **3810448**, HOW CAN I HELP YOU?

IT'S ABOUT MEASURING OUR **CONFIDENCE.**

NEVERTHELESS, STATISTICS **DOES** INVOLVE **WRESTLING WITH NUMBERS...**

WHICH ONE WOULD YOU **RATHER FIGHT?**

I HAVE WON **147** MATCHES AND LOST ONLY **17.**

I AM VICTOR **89.6%** OF TIME!

...AND THAT'S **NOT ALWAYS EASY.**

ERM... I'M NOT FEELING VERY **CONFIDENT** AT THE MOMENT.

SOME NUMBERS **SEEM SCARY**...

YOU HAVE MORE THAN **TWO POUNDS** OF BACTERIA LIVING INSIDE YOU!

...OTHERS **SEEM REASSURING**.

IF THEY'VE SOLD MORE THAN A **BILLION** HAMBURGERS...

...THEY **MUST** BE GOOD!

SOME **REVEAL A GREAT DEAL**...

WE'VE ERADICATED **99.99%** OF THE **SMALLPOX VIRUS** WORLDWIDE.

...OTHERS, **NOT SO MUCH**...

OUR CALCULATIONS **PROVE** THAT THE WORLD WILL END ON **FEBRUARY 29, 2024!**

...AND IT CAN BE **HARD TO TELL THE DIFFERENCE**.

I HEARD THAT **12.4%** OF ALL SODA DRINKERS DIE EVERY DAY!

WHAT DOES THAT EVEN **MEAN**?

ALL THESE **FACTS ABOUT NUMBERS** MAKE IT **EASY TO USE THEM...**

...TO **LIE.**

IF YOU WEAR THIS TIE...

...EVERYONE WILL THINK YOU'RE **POWERFUL.**

AND IF I MENTION A NUMBER...

...EVERYONE WILL THINK I'M **SMART.**

SADLY, THIS CAN MAKE PEOPLE **OVERLY SUSPICIOUS ABOUT NUMBERS GENERALLY...**

...AND **IGNORANT** OF THEIR **VERY REAL POWER.**

I DON'T CARE IF WE PUMP 5.5 MILLION THOUSAND METRIC TONS OF CO_2 INTO THE AIR...

...**THAT'S JUST A NUMBER.**

WITHOUT NUMBERS THERE WOULD BE **NO VIDEO GAMES...**

...AND YOU WOULDN'T BE ABLE TO **BUY STUFF.**

THE **SOLUTION** TO THIS PROBLEM...

SOME NUMBERS ARE **TRUE**...

...BUT YOU MUST **RESIST THE TEMPTATION** TO BELIEVE THE FALSE ONES!

HOW DO I KNOW **WHICH IS WHICH?**

...IS TO TREAT **ALL NUMBERS**...

...NO MATTER HOW **LARGE**...

...OR **SMALL**...

...OR **SLEEP INDUCING**...

ZZZZZ

...WITH A **HEALTHY DOSE OF SKEPTICISM**.

THESE COOKIES ARE **100%** ORGANIC, AND **98.3%** VEGAN...

...BUT THEIR TRANS-ISOMER LACTO-OVO QUOTIENT IS **WELL WITHIN** THE RECOMMENDED HOMOCYSTEINE RATIO.

I'LL TAKE ONE **WITH A GRAIN OF SALT**, PLEASE.

THAT'S THE **FIRST LESSON OF THIS BOOK**.

HELP!

I'M **UNCERTAIN** ABOUT THESE NUMBERS.

GOOD!

LEARN TO **ENJOY IT!**

IN STATISTICS, YOU'RE **SUPPOSED** TO FEEL THAT WAY.

SO WHETHER YOU'RE SOMEBODY WHO FEELS **COMFORTABLE AROUND NUMBERS**...

...OR **NOT**...

...YOU SHOULD **ALWAYS ASK THESE QUESTIONS** OF **ANY NUMBER YOU ENCOUNTER:**

WHERE DID YOU COME FROM?

WHO MADE YOU?

AND WHY?

CHAPTER 2
RANDOM RAW DATA

SINCE THE **DAWN OF CIVILIZATION**...

...PEOPLE HAVE HAD THE **URGE TO COUNT THINGS.**

GOOD MORNING, *I AM YOUR LEADER!*

BUT **WHY?**

BECAUSE HE'S STRANGLED **764 MEN!**

IN FACT, THE EARLIEST **FORMS OF WRITING** WERE **INVENTED FOR COUNTING.**

HOW DO I KNOW IF I HAVE **ENOUGH OXEN** AND **GRAIN** TO FEED **MY PEOPLE?**

...OR **ENOUGH SOLDIERS TO FIGHT MY ENEMIES?**

I KNOW! LET'S **KEEP TRACK OF THOSE THINGS**... ...WITH LITTLE **DRAWINGS LIKE THESE.**

AS CIVILIZATIONS **GREW**...

...SO DID THE **NUMBER OF THINGS** NEEDING TO BE COUNTED.

BY MY CALCULATIONS, YOUR EMPIRE STRETCHES **TO THE ENDS OF THE EARTH.**

GREAT, SO HOW MANY **OXTAILS** CAN WE COOK FOR MY BIRTHDAY PARTY NEXT WEEK?

BUT THIS CREATED A **NEW CHALLENGE**.

TELL ME THE NUMBER OF **ENEMIES EACH OF MY SOLDIERS HAS STRANGLED!**

SOMETIMES IT'S **IMPOSSIBLE** TO COUNT **ALL** THE THINGS YOU WANT TO KNOW ABOUT.

THERE'S, LIKE, A **GAZILLION** SOLDIERS...

...I CAN'T POSSIBLY TALK TO **ALL** OF THEM!

YOUR PROBLEM, NOT HIS.

WHICH IS WHY, LONG AGO, SOMEONE DREAMED UP THE **STRATEGY**...

...OF STUDYING A **SAMPLE**...

...TO LEARN SOMETHING ABOUT AN ENTIRE **POPULATION**.

I KNOW! I WILL TALK TO **SOME** OF THE SOLDIERS...

...AND THEN I'LL USE WHAT I LEARN TO MAKE A GUESS ABOUT...

...**ALL** OF THE SOLDIERS!

USING SAMPLES TO DESCRIBE POPULATIONS IS **CLEVER**...

...BUT THERE ARE A FEW FACTS TO KEEP IN MIND BEFORE WE **ACTUALLY TRY TO START DOING IT.**

JUST BECAUSE **I DON'T KNOW EVERYTHING**...

...DOESN'T MEAN I KNOW **NOTHING!**

FIRST, IT'S **IMPOSSIBLE** TO USE A SAMPLE TO ACHIEVE **ABSOLUTE CERTAINTY** ABOUT A POPULATION.

IF YOU WANT TO KNOW **THE WHOLE TRUTH** ABOUT **ALL THESE MOSQUITOES**...

...YOU HAVE TO MEASURE **ALL** THESE MOSQUITOES.

THAT'S WHY STATISTICS IS ABOUT MAKING OUR **BEST POSSIBLE GUESS**...

...AND **NEVER** ABOUT **BEING CERTAIN.**

LET'S MEASURE **100 OF THEM**...

...AND SEE WHAT THAT SUGGESTS ABOUT THE REST.

GOOD PLAN.

YOU'LL NEVER BE CERTAIN, BUT **AT LEAST YOU WON'T BE EATEN ALIVE!**

28

SECOND, IF WE'RE **STUCK WITH A SINGLE SAMPLE**...

I CAN LEARN TO SAY SOMETHING ABOUT **ALL THE SQUID IN THE OCEAN**...

...BY STUDYING **ONLY THESE 35!**

...WE'D BETTER BE SURE WE **COLLECTED IT CAREFULLY!**

UM, DID YOU **WASH YOUR HANDS** BEFORE YOU PICKED THOSE UP?

BECAUSE **ANY MISTAKES** WE MAKE WHEN WE COLLECT OUR SAMPLE...

...CAN **TOTALLY SCREW UP** WHAT WE CONCLUDE ABOUT THE LARGER POPULATION.

YOU LEFT YOUR **COFFEE ON THE SCALE.**

AND THIS ONE IS AN **OCTOPUS.**

YOU CAN USE EITHER **INCHES** OR **CENTIMETERS**, BUT **NOT BOTH.**

WHOSE HAIR IS THIS?

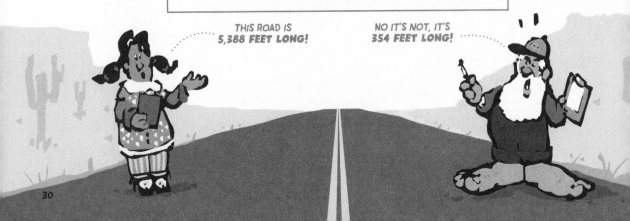

COLLECTING SAMPLES IS ALSO TOUGH WHEN WE'RE TRYING TO GET A SENSE OF **WHAT PEOPLE THINK...**

WHICH COLOR WORKS BETTER, **RED** OR **GREEN?**

I'M COLOR-BLIND.

...OR **HOW THEY FEEL...**

HOW MUCH DO YOU **LOVE YOUR NEIGHBORS?**

I HATE 'EM!

...OR ANYTHING ELSE THEY **MIGHT NOT WANT TO TALK ABOUT...**

HOW MANY **BANKS HAVE YOU ROBBED?**

DEPENDS WHO'S ASKING.

...OR **MIGHT WANT TO EXAGGERATE.**

THIS SOLDIER TOLD ME HE'D PERSONALLY **STRANGLED 765 MEN!**

AND YOU **BELIEVED** HIM?

PERHAPS THE **BIGGEST CHALLENGE** IN COLLECTING A SAMPLE...

I'M GOING TO INTERVIEW 100 SOLDIERS!

BUT WHICH 100 SOLDIERS SHOULD I CHOOSE?

...IS FIGURING OUT EXACTLY **WHAT TO INCLUDE IN IT.**

SHOULD I CHOOSE THE **POLITE** SOLDIERS WHO ARE HANGING OUT AT THE COFFEE SHOP...

...OR THE **TERRIFYING** ONES WHO I CAN FIND AT THE GYM?

THE GOAL IS TO AVOID ANY **BIAS IN OUR SAMPLE**...

...THAT MIGHT LEAD US TO **MISCHARACTERIZE THE POPULATION.**

IF YOU INTERVIEW TOO MANY **POLITE** SOLDIERS...

...YOU'RE GOING TO THINK THE ARMY IS **NICER THAN IT REALLY IS!**

IF YOU INTERVIEW TOO MANY **TERRIFYING** SOLDIERS...

...YOU'RE GOING TO THINK THE ARMY IS **SCARIER THAN IT REALLY IS!**

IDEALLY, WE'D LIKE TO GATHER A SAMPLE THAT **ACCURATELY MIRRORS THE POPULATION.**

THIS CAN SEEM LIKE AN **IMPOSSIBLE TASK...**

...BUT STATISTICIANS HAVE CONCOCTED **A RELIABLE WAY AROUND IT.**

TO AVOID BIAS, WE **ALWAYS** COLLECT SAMPLES **RANDOMLY.**

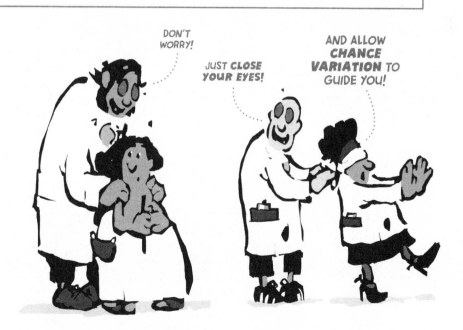

COLLECTING RANDOM SAMPLES IS A **SIMPLE IDEA**...

LET'S PUT THE ENTIRE ARMY **IN THIS HELMET**...

...AND RANDOMLY **PULL OUT ONE SOLDIER AT A TIME.**

IT'S LIKE **BINGO!**

... THAT CAN BE **HARD TO PULL OFF.**

CLIMB IN, BOYS!

IN PRACTICE, WE OFTEN HAVE TO IMAGINE **ALL THE THINGS** THAT **MIGHT BIAS** OUR SAMPLE...

I DON'T WANT TO INTERVIEW TOO MANY SOLDIERS WHO ARE **ASLEEP IN THE GUTTER**...

...OR WHO **STINK**...

...OR WHO **LIVE IN THIS CITY**...

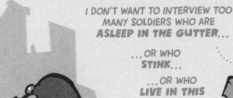
... AND **MAKE SURE THEY DON'T.**

SO I'LL **CLOSE MY EYES**...

...AND **HOLD MY NOSE**...

...AND **WANDER AROUND THE ENTIRE EMPIRE**...

...AND INTERVIEW THE SOLDIERS I **RANDOMLY BUMP INTO.**

OF COURSE, EVEN WHEN WE COLLECT OUR SAMPLE **RANDOMLY**...

BLINDFOLDS **ON!**

... WE STILL CAN'T **GUARANTEE** THAT IT WILL **ACCURATELY MIRROR** THE POPULATION IT COMES FROM.

IN FACT, ANY ONE **RANDOM SAMPLE**...

... WILL **PROBABLY LOOK DIFFERENT** FROM **THE OVERALL POPULATION**...

...AND FROM **ANY OTHER** RANDOM SAMPLE WE MIGHT RANDOMLY COLLECT.

I GRABBED THESE 100 FLEAS RANDOMLY.

THERE ARE A GAZILLION FLEAS ON THIS DOG.

HERE ARE 100 OTHER RANDOM FLEAS.

THE REASON RANDOM SAMPLING WORKS SO WELL IS THAT IT MEANS WE'RE **JUST AS LIKELY** TO GRAB ANY ONE SAMPLE...

...AS ANY OTHER...

HERE ARE 100 RANDOM **PORCUPINES**.

HERE ARE **ANOTHER** 100 RANDOM PORCUPINES.

...AND **IF** THEY'RE DIFFERENT...

...IT'S **ONLY** BECAUSE OF **CHANCE**.

THESE GUYS HAVE **LONGER QUILLS** THAN THOSE GUYS...

...HOW **RANDOM**.

I'M STUDYING **ALL** THE FISH IN THE SEA...

...AND I NEED TO COLLECT THEM **RANDOMLY.**

SO TODAY I'M GRABBING ONE THAT FOR SOME REASON LIVES **RIGHT HERE.**

...BUT IT'S **INCREDIBLY IMPORTANT** TO GET IT RIGHT...

IT'S WORTH **WALKING TO THE ENDS OF THE EARTH**...

...OR **DIVING TO THE BOTTOM OF THE SEA!**

...BECAUSE **RANDOM SAMPLING***
IS THE **KEY** TO ALL STATISTICAL INQUIRY.

IF **THESE** FISH AREN'T **RANDOM**...

...WE CAN'T SAY **ANYTHING** ABOUT THOSE FISH.

* SEE **PAGE 214** FOR SOME TECHNICAL DETAILS.

IN THIS CHAPTER, WE'VE LEARNED HOW **RANDOM SAMPLING** CAN HELP US **AVOID BIAS**.

I COULD BE **SCREWING UP MY IMAGE OF THE ARMY**...

...BY INTERVIEWING ONLY **SOLDIERS WHO WON'T KILL ME**.

BUT RANDOM SAMPLES ARE **ALSO** A VITAL PART OF THE **STATISTICAL MACHINERY** WE'LL BE LEARNING ABOUT LATER.

ALL THE TOOLS WE'LL LEARN ABOUT IN PART TWO **REQUIRE RANDOM SAMPLES**.

YOU PUT YOUR RANDOM SAMPLE IN HERE...

...**ADJUST THIS KNOB**...

...AND OUT POPS **A CONFIDENCE INTERVAL!**

IF YOUR SAMPLE **ISN'T RANDOM**...

...THE ONLY THING THAT POPS OUT IS **GOBBLEDYGOOK!**

CHAPTER 3
SORTING

HERE ARE
**50 RANDOM
RHINOS...**

...WHEN WE'RE **CURIOUS TO KNOW SOMETHING** ABOUT THE LARGER POPULATION IT COMES FROM.

WELL...

WHAT DO YOU WANT TO KNOW ABOUT US?

SOMETIMES WE'RE CURIOUS ABOUT **QUALITIES** OR **CATEGORIES**...

...AND SOMETIMES WE'RE CURIOUS ABOUT QUESTIONS WE CAN ANSWER WITH **NUMBERS**.

DO YOU **USE DEODORANT**?

WHERE WERE YOU **BORN**?

WHAT'S YOUR **FAVORITE FOOD**?

HOW MUCH DO YOU **SLEEP**?

HOW THICK IS YOUR **SKIN**?

HOW OFTEN DO YOU **BATHE**?

THE DISTINCTION IS IMPORTANT BECAUSE THE **KIND OF QUESTION** WE ASK...

WHICH **TYPE** OF SHOE DO YOU PREFER?

WHAT **SIZE** SHOE DO YOU WEAR?

...DETERMINES WHETHER WE END UP WITH **CATEGORICAL DATA**...

...OR **NUMERICAL** DATA...

I PREFER **BOOTS**.

PUMPS, BABY, ALL THE WAY.

I'M DEFINITELY A **FLIP-FLOP** KIND OF GUY.

YOUR SHOE SIZE IS 14.

MINE IS 16.5!

12.

...AND THE TWO KINDS OF DATA **DON'T MIX**.

THEY'RE LIKE **OIL** AND **WATER**.

WE GATHER **CATEGORICAL DATA**...

... WHEN WE'RE STUDYING FEATURES THAT WE CAN **DESCRIBE ONLY WITH WORDS**,...

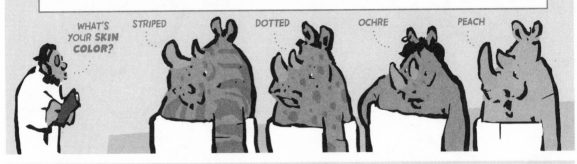

WHAT'S YOUR **SKIN COLOR?**

STRIPED

DOTTED

OCHRE

PEACH

WHICH PARTY DO YOU VOTE FOR?

LIBERTARIAN

COMMUNIST

REPUBLICAN

INDEPENDENT

... OR **YES/NO ANSWERS**,

ARE YOU PERSONALLY **ATTRACTIVE?**

YES

YES

I THINK SO.

ARE YOU **KIDDING ME?**

AFTER WE GATHER CATEGORICAL DATA WE CAN EASILY **PILE IT**...

... OR **SLICE IT**...

MORE SAMPLE RHINOS PREFER **THISTLES** THAN **STONES!**

NEEDS KETCHUP.

MOST SAMPLE RHINOS FEEL **OPTIMISTIC!**

Number of Rhinos

Chewing Preference

... TO GIVE US A SENSE OF THE **PROPORTIONS** IN OUR SAMPLE.

... WHEN WE'RE STUDYING FEATURES THAT WE CAN **COMPARE USING NUMBERS.**

HOW OLD ARE YOU, IN YEARS?

10.3   2.9   16.9   829.1

HOW LONG IS YOUR HORN, IN CENTIMETERS?

43.3   37.2   53.5   6.4

HOW BAD IS YOUR EYESIGHT?

20/80   20/900   20/400   20/2,400

AS WE'LL SEE IN PART TWO, ALL THESE NUMBERS MAKE NUMERICAL DATA **MUCH MORE USEFUL** OVERALL.

YOU PUT YOUR RANDOM NUMBERS IN HERE...

...SLIDE THIS THINGY OVER...

...AND OUT POPS **A P-VALUE!**

THE **CRUCIAL DIFFERENCE**
BETWEEN THE TWO TYPES OF DATA...

NOT ALL DATA
ARE CREATED
EQUAL.

...IS THAT **WE CAN'T DO MATH
ON CATEGORICAL DATA...**

WHAT'S THE
AVERAGE COLOR
IN YOUR SAMPLE?

ERM... SORT OF
AQUAMARINE OCHRE
PASTY-ISH?

...BUT WE **CAN DO MATH
ON NUMERICAL DATA!**

WHAT'S THE
**AVERAGE
LENGTH** IN YOUR
SAMPLE?

0.004 METERS
ON THE NOSE!

THIS FACT MAKES NUMERICAL DATA **EXCITING TO STATISTICIANS...**

SD EQUALS
SIGMA OVER
THE SQUARE
ROOT OF **n**!

I LOVE IT WHEN
YOU **TALK MATH
TO ME.**

...BUT **INTIMIDATING TO NORMAL PEOPLE.**

SD EQUALS
SIGMA OVER
THE SQUARE
ROOT OF **n**!

AAHHHH!

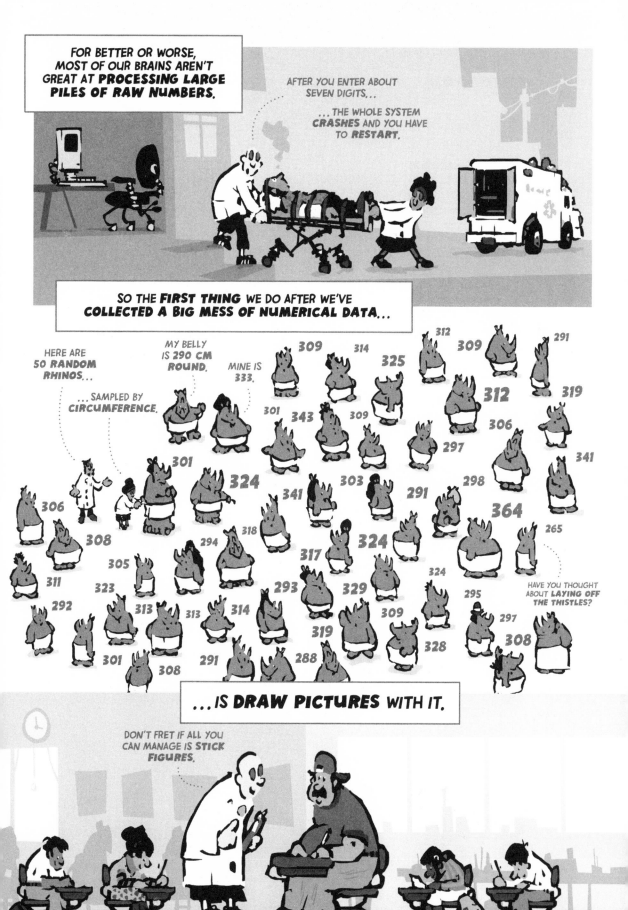

THE MOST BASIC PICTURE OF NUMERICAL DATA IS CALLED A **HISTOGRAM**.

TO DRAW A **HISTOGRAM** OF OUR SAMPLE...

...WE START WITH **A NUMBER LINE**.

THESE NUMBERS ON THE GROUND RANGE BETWEEN THE **SMALLEST**...

...AND **LARGEST** VALUES WE OBSERVED.

265

364

Girth (in cm) 270 280 290 300 310 320 330 340 350 360 370

THEN WE **PILE OUR DATA ON TOP OF IT**...

...**PIECE BY PIECE**.

A HISTOGRAM IS LIKE A **BIG PILE OF BOXES**.

EACH RHINO GETS ASSIGNED TO **ONE BOX**.

THIS RHINO'S BELLY IS 343 CM AROUND...

...SO SHE BELONGS **RIGHT HERE**.

10
9
8
7
6
5
4

Number of Rhinos

Girth (in cm) 270 280 290 300 310 320 330 340 350 360 370

ANOTHER USEFUL WAY TO VISUALIZE NUMERICAL DATA IS WITH A **BOXPLOT.**

TO DRAW A **BOXPLOT** OF OUR SAMPLE...

... WE START WITH **THE SAME NUMBER LINE...**

I'M THE **SMALLEST VALUE** IN THE ENTIRE SAMPLE.

I'M THE **LARGEST VALUE** IN THE ENTIRE SAMPLE.

265

364

Girth (in cm)   270   280   290   300   310   320   330   340   350   360   370

... BUT IN THIS CASE WE **CRAM THE MIDDLE 50%** OF OUR SAMPLE VALUES INTO **ONE BIG BOX.**

THIS BOX GIVES US A SENSE OF WHERE THE **BULK OF THE DATA SITS...**

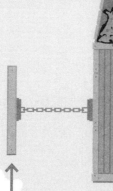

THEN WE INDICATE THE **MINIMUM...**

... **MIDDLE...**

... AND **MAXIMUM** INDIVIDUAL VALUES WITH THESE BARS.

THAT'S ME.

THAT'S ME.

THAT'S ME.

265

312

364

Girth (in cm)   270   280   290   300   310   320   330   340   350   360   370

... THAT INCLUDES **PRECISE DETAILS.**

IT'S LIKE A **MOUNTAIN RANGE!**

WE CAN USE IT TO EXPLORE THE **PEAKS**...

... AND **VALLEYS.**

FOR EXAMPLE, THIS HISTOGRAM OF **RHINO HORN LENGTH**...

49 OF US HAVE HORNS THAT ARE BETWEEN 5 AND 55 CM LONG...

... BUT MINE IS 97 CM!

Number of Rhinos

Horn Length (in cm)   0   10   20   30   40   50   60   70   80   90   100

... CLEARLY SHOWS THAT **ONE RHINO** IS **MUCH HORNIER** THAN THE OTHERS.

48

ON THE OTHER HAND, **BOXPLOTS** CAN BE ESPECIALLY USEFUL WHEN WE WANT AN **OVERVIEW** OF OUR DATA...

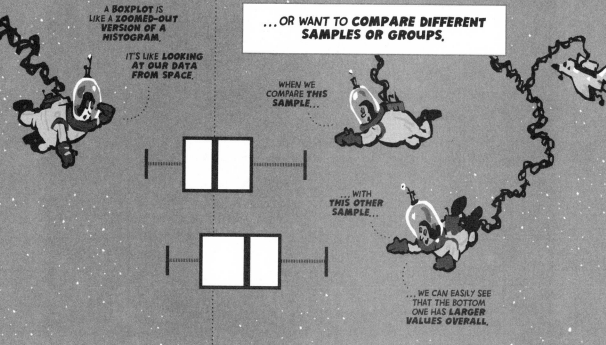

A BOXPLOT IS LIKE A **ZOOMED-OUT** VERSION OF A HISTOGRAM.

IT'S LIKE **LOOKING** AT OUR DATA FROM SPACE.

...OR WANT TO **COMPARE DIFFERENT** SAMPLES OR GROUPS.

WHEN WE COMPARE **THIS** SAMPLE...

...WITH **THIS OTHER** SAMPLE...

...WE CAN EASILY SEE THAT THE BOTTOM ONE HAS **LARGER VALUES OVERALL.**

BOXPLOTS CAN GIVE US A QUICK SENSE OF **HOW DATA CLUMPS TOGETHER...**

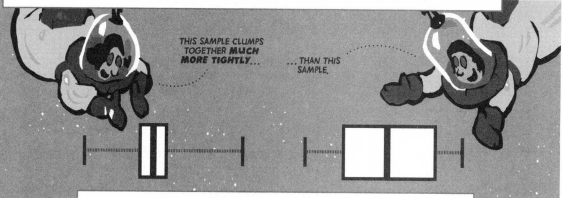

THIS SAMPLE CLUMPS TOGETHER **MUCH MORE TIGHTLY...**

...THAN THIS SAMPLE.

...AND WHETHER IT **TRAILS OFF** IN ONE DIRECTION OR THE OTHER.

HMMM, THE BULK OF THIS DATA IS **WAY OFF TO THE LEFT...**

...BUT THE BULK OF **THIS DATA IS WAY OFF TO THE RIGHT!**

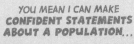

IT MAY SEEM SURPRISING THAT STATISTICS INVOLVES **DRAWING PICTURES**.

YOU MEAN I CAN MAKE CONFIDENT STATEMENTS ABOUT A POPULATION...

...USING **ONLY** THIS DOODLE?

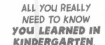

ALL YOU REALLY NEED TO KNOW **YOU LEARNED IN KINDERGARTEN**.

BUT THE FACT IS, THE **FIRST THING** WE SHOULD **ALWAYS** DO WITH OUR DATA IS **LOOK AT IT**.

YOU'D BE **AMAZED** HOW OFTEN PEOPLE FORGET THIS.

BECAUSE ALTHOUGH WE MIGHT BE **TEMPTED** BY **MORE SOPHISTICATED MATHEMATICAL TOOLS**...

HEY, BUD, WANNA BUY A **NON-PARAMETRIC BAYESIAN HIERARCHICAL ALGORITHM**?

...SIMPLE PICTURES KEEP US FOCUSED ON WHAT THE DATA **ACTUALLY HAVE TO SAY**.

BEFORE YOU GENERATE AN **AUTHORITATIVE** SOUNDING NUMBER...

...**DRAW A PICTURE!**

A HISTOGRAM IS WORTH A THOUSAND P-VALUES.

CHAPTER 4
DETECTIVE WORK

*THAT'S NICE, BUT WHAT DOES IT **MEAN?***

ANALYZING DATA IS LIKE **SOLVING A MYSTERY**.

IT WAS PROFESSOR PLUM...

...WITH THE **CANDLESTICK**...

...IN THE **LOUNGE**...

...WHERE HE **FLEW INTO A RAGE** WHILE **TRYING TO LEARN STATISTICS**!

OUR ULTIMATE GOAL IS TO **GATHER EVIDENCE** FROM **ONE RANDOM SAMPLE**...

...AND USE IT TO PIECE TOGETHER **A STORY ABOUT A POPULATION**.

GIVE ME **ONE GROUP OF RANDOM SUPERVILLAINS**...

...AND I CAN CONFIDENTLY TELL YOU SOMETHING ABOUT **ALL THE SUPERVILLAINS**...

...ALL THE SUPERVILLAINS **IN THE ENTIRE WORLD!**

BUT **FIRST** WE NEED TO LEARN HOW TO DO SOME **BASIC DETECTIVE WORK**.

WHEN WE **START TO INVESTIGATE** ANY PILE OF DATA...

THIS HISTOGRAM SHOWS 64 RANDOM SUPERVILLAINS SORTED BY **AGE.**

Supervillain count

Age | 10 | 20 | 30 | 40 | 50 | 60 | 70 | 80 | 90 | 100

...WE ALWAYS LOOK AT **FOUR PRIMARY CHARACTERISTICS**...

SAMPLE SIZE	**SHAPE**	**LOCATION**	**SPREAD**
HOW **MUCH** DATA IS IN THERE?	WHAT DOES THE PILE **LOOK LIKE**?	WHERE IS IT, EXACTLY?	HOW **WIDE** IS IT?

...AND WE'RE GOING TO SPEND THIS CHAPTER **LEARNING ABOUT THEM.**

WHAT MYSTERIES ARE HIDDEN WITHIN THIS **MOUND OF MURDERERS?**

LET'S **SIFT** FOR CLUES.

SAMPLE SIZE

HOW **MANY** PIECES OF DATA ARE IN THERE?

SAMPLE SIZE* IS THE **FIRST THING** TO LOOK FOR IN ANY PILE OF DATA...

HOW **MANY** RANDOM SUPERVILLAINS DID WE **PILE UP?**

64

...AND IT'S EASY TO SEE **WHY IT MATTERS.**

IF YOU HAD ONLY **A FEW** RANDOM VILLAINS IN YOUR SAMPLE...

LIKE US FIVE!

...YOU COULDN'T SAY MUCH OF **ANYTHING** ABOUT OUR OVERALL POPULATION.

SORRY, THIS PICTURE OF YOUR DATA **DOESN'T REALLY HELP MUCH.**

* SEE **PAGE 214.**

IN GENERAL, A **LARGER SAMPLE SIZE** IS **BETTER!**

WITH ONLY A FEW DATA POINTS WE **CAN'T SEE MUCH...**

...BUT AS WE PILE UP **MORE RANDOM DATA...**

...OUR PICTURE BECOMES **MORE HELPFUL!**

AS WE'LL DISCOVER LATER, THE SIZE OF A SAMPLE IS **DIRECTLY RELATED** TO THE **LEVEL OF CONFIDENCE** WE CAN HAVE ABOUT A POPULATION.

SIZE MATTERS!

IF WE PUT IN **MORE RANDOM DATA...**

...WE GET **MORE CONFIDENCE!**

UNFORTUNATELY, IN PRACTICE THE SIZE OF A SAMPLE IS **ALWAYS LIMITED BY SOMETHING.**

WE **CAN'T HAVE TOO MANY** RANDOM VILLAINS, WATSON!

I'D BE HAPPY TO COLLECT MORE, HOLMES...

...BUT WE **RAN OUT OF HANDCUFFS.**

SHAPE

THE SHAPE OF
EACH SAMPLE IS
UNIQUE...

...LIKE A
FINGERPRINT!

THE MOMENT WHEN SOMEONE DISCOVERS **THE SHAPE**
OF THEIR SAMPLE CAN BE **EXTREMELY SUSPENSEFUL**...

I'M SORRY, MRS. JONES,
YOUR DATA **ISN'T**
NORMAL.

MERCY!

...BECAUSE HOWEVER A PILE OF DATA IS SHAPED,
IT'S **ALWAYS SHAPED THAT WAY FOR A REASON**.

YOUR SAMPLE
LOOKS **LIKE A**
CAMEL...

...IT MUST BE
BECAUSE OF SOME
UNDERLYING
CONDITION.

WE CALL A PILE OF DATA **NORMAL** WHEN SOMETHING IS CAUSING IT TO **CLUMP AROUND ONE PARTICULAR VALUE.**

WE CALL A PILE OF DATA **SKEWED** WHEN SOMETHING IS CAUSING IT TO **TRAIL OFF MORE IN ONE DIRECTION THAN THE OTHER.**

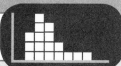

LOCATION

WHERE DOES THE DATA **CLUSTER?**

LOCATION IS A MEASURE OF **WHERE THE BULK OF THE DATA SITS** ON A NUMBER LINE.

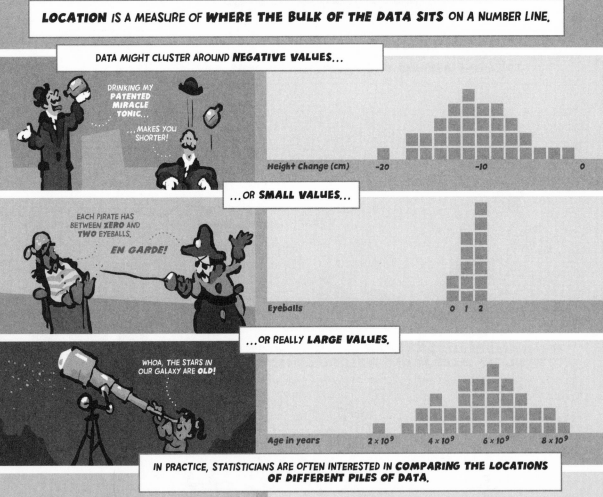

DATA MIGHT CLUSTER AROUND **NEGATIVE VALUES**...

DRINKING MY **PATENTED MIRACLE TONIC**...

...MAKES YOU SHORTER!

Height Change (cm) -20   -10   0

...OR **SMALL VALUES**...

EACH PIRATE HAS BETWEEN **ZERO** AND **TWO** EYEBALLS.

EN GARDE!

Eyeballs   0 1 2

...OR REALLY **LARGE VALUES**.

WHOA, THE STARS IN OUR GALAXY ARE **OLD!**

Age in years 2×10^9 4×10^9 6×10^9 8×10^9

IN PRACTICE, STATISTICIANS ARE OFTEN INTERESTED IN **COMPARING THE LOCATIONS OF DIFFERENT PILES OF DATA.**

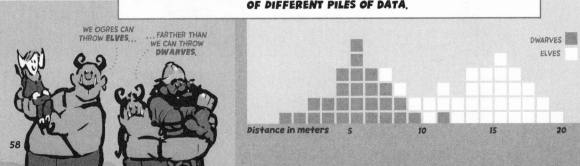

WE OGRES CAN THROW **ELVES**...   ...FARTHER THAN WE CAN THROW **DWARVES**.

DWARVES
ELVES

Distance in meters   5   10   15   20

DEFINING LOCATION WITH WORDS CAN BE **TRICKY**...

IT'S WHERE THE DATA **CLUSTERS**!

SO IT'S KIND OF A **BLOB**?

IT'S THE **CENTRAL TENDENCY.**

THAT'S STILL A BIT **VAGUE**...

...CAN YOU BE MORE **SPECIFIC**?

...SO WE OFTEN DESCRIBE IT WITH A SINGLE NUMBER: **THE AVERAGE.***

*TECHNICALLY, THIS NUMBER IS ALSO CALLED **THE MEAN**. TO LEARN HOW TO CALCULATE IT, TURN TO **PAGE 214**.

WHEN YOU REACH **THIS EXACT NUMBER**, STOP AND YOU'RE THERE!

NOW WE'RE **TALKIN'.**

TO CALCULATE THE AVERAGE, WE SIMPLY ADD UP ALL THE DATA VALUES...

...THEN DIVIDE BY THE NUMBER OF DATA VALUES.

PIRATES, THROW YOUR YEAR'S SALARY INTO THIS BUCKET!

ARGH.

THE TOTAL NUMBER OF DOUBLOONS IS 6,000...

...AND THERE BE **50 PIRATES.**

SO THE **AVERAGE INCOME** ON THIS PIRATE SHIP IS **120 DOUBLOONS PER YEAR!**

HOWEVER, ALTHOUGH THE AVERAGE IS USEFUL AND PRECISE AS A MEASURE OF LOCATION, IT ISN'T PERFECT.

WHOA, ON AVERAGE, WE'RE **RICH!**

THEN HOW COME ONE-EYED JACK **CAN'T** AFFORD A BETTER EYEBALL?

UNFORTUNATELY, **AVERAGES CAN BE DECEPTIVE.**

JUST BECAUSE OUR AVERAGE YEARLY INCOME IS 120 DOUBLOONS...

...DOESN'T MEAN THAT **MOST OF US ARE RICH!**

FOR EXAMPLE, IF A PILE OF DATA IS **SKEWED...**

...AN AVERAGE VALUE **CAN BE SERIOUSLY MISLEADING.**

MOST PIRATES ON THIS SHIP...

...EARN MUCH LESS THAN THE AVERAGE.

THE AVERAGE IS SO HIGH ONLY BECAUSE **GREENBEARD** IS ON BOARD.

Avg

Pirate Count

Income | 0 | 100 | 200 | 300 | 400 | 500

WITH SKEWED DATA, THE **MEDIAN** IS OFTEN MORE REVEALING AS A MEASURE OF LOCATION...

THE MEDIAN IS THE EXACT **MIDDLE VALUE!**

THERE ARE THE **SAME NUMBER OF DATA POINTS** TO ITS LEFT AND TO ITS RIGHT.

median = 72

IT'S EASY TO SEE ON A **BOXPLOT.**

...BECAUSE IT CAN GIVE A BETTER SENSE OF A **"TYPICAL"** VALUE.

I'M THE **MEDIAN.**

I MAKE 72 DOUBLOONS A YEAR.

I'M **AVERAGE.**

I MAKE **120!**

I MAKE 500...

...HAR HAR.

I'M THE REASON THOSE TWO ARE **SO FAR APART.**

Income | 0 | 100 | 200 | 300 | 400 | 500

THE AVERAGE PIRATE HAS *1.28 EYEBALLS*...

...EARNS *120 DOUBLOONS*...

...AND DRINKS *82.9 LITERS* OF GROG EACH YEAR.

I'LL TAKE ALL THAT **WITH A GRAIN OF SALT**, PLEASE.

...IT'S IMPORTANT TO REMEMBER THAT AN AVERAGE TELLS US ONLY **ONE VERY PRECISE THING** ABOUT OUR DATA.

THE SUM OF ALL THE MEASUREMENTS...

...DIVIDED BY THE NUMBER OF MEASUREMENTS.

THAT'S ALL!

WHICH IS ONE REASON WE SHOULD **NEVER** THINK ABOUT THE **LOCATION** OF ANY PILE OF DATA ...

LOOK, HOLMES, THE **AVERAGE SUPERVILLAIN** IN OUR SAMPLE SCORED 510 ON THE MATH SAT!

... WITHOUT ALSO THINKING ABOUT ITS **SHAPE**...

BUT WATSON, BECAUSE **THIS BIG CLUMP** OF THEM SUCKS AT MATH...

...ALMOST **NONE OF THEM SCORED ANYWHERE NEAR THE AVERAGE VALUE.**

...AND THIS **CLUMP** OF THEM ROCKS AT MATH...

Avg

Supervillain Count

Math SAT Score

400

600

800

... AND ABOUT ITS **SPREAD**, WHICH IS COMING UP NEXT.

SPREAD

HOW WIDE IS IT?

SPREAD IS A MEASURE OF THE **WIDTH OF A PILE OF DATA**...

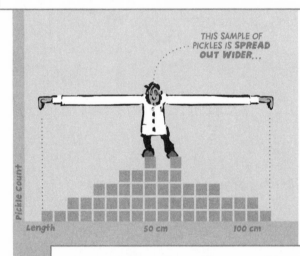

THIS SAMPLE OF PICKLES IS **SPREAD OUT WIDER**...

Pickle Count

Length | 50 cm | 100 cm

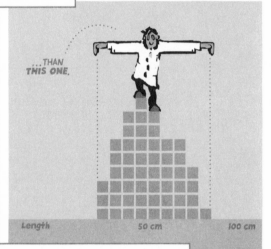

...THAN **THIS ONE.**

Length | 50 cm | 100 cm

...BUT IT'S ALSO A MEASURE OF **VARIATION.**

THIS SAMPLE OF PICKLES HAS **MORE VARIATION** IN IT...

...THAN **THIS ONE.**

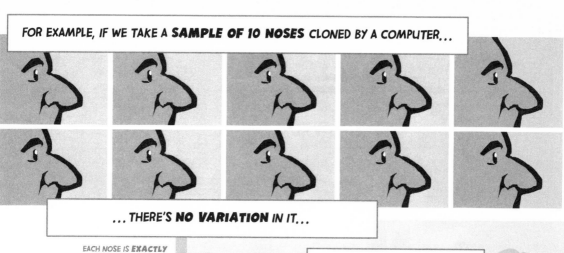

FOR EXAMPLE, IF WE TAKE A **SAMPLE OF 10 NOSES** CLONED BY A COMPUTER...

... THERE'S **NO VARIATION** IN IT...

EACH NOSE IS **EXACTLY 0.23 CM LONG**.

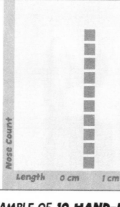

... AND THUS **NO SPREAD**.

BORING!

Nose Count

Length   0 cm   1 cm   10 cm   20 cm

HOWEVER, IF WE TAKE A SAMPLE OF **10 HAND-DRAWN NOSES**...

... THERE'S A **GREAT AMOUNT OF VARIATION** IN IT...

THESE HAND-DRAWN NOSES VARY BETWEEN **0.1 CM** SHORT...

... AND **16.98 CM** LONG.

... AND THUS A FAIRLY **WIDE SPREAD**.

WIDER SPREAD EQUALS MORE VARIATION!

Nose Count

Length   0 cm   1 cm   10 cm   20 cm

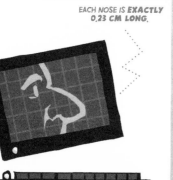

63

ONE STRAIGHTFORWARD WAY TO **MEASURE SPREAD** IS TO TAKE THE **OVERALL RANGE**...

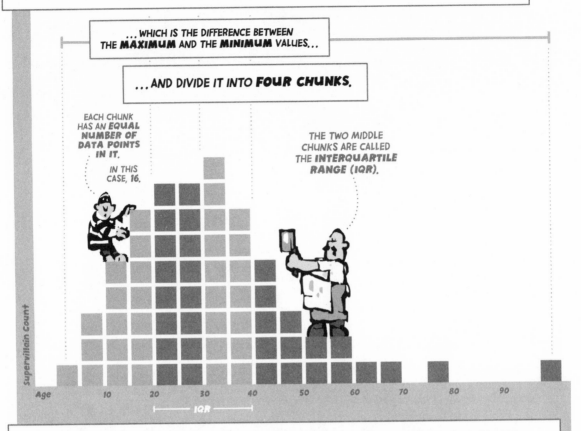

...WHICH IS THE DIFFERENCE BETWEEN THE **MAXIMUM** AND THE **MINIMUM** VALUES...

...AND DIVIDE IT INTO **FOUR CHUNKS**.

EACH CHUNK HAS AN **EQUAL NUMBER OF DATA POINTS IN IT.**

IN THIS CASE, 16.

THE TWO MIDDLE CHUNKS ARE CALLED THE **INTERQUARTILE RANGE (IQR).**

Supervillain Count

Age   10   20   30   40   50   60   70   80   90

IQR

THIS GIVES US A SENSE OF THE VARIATION **WITHIN EACH PART** OF THE OVERALL SAMPLE...

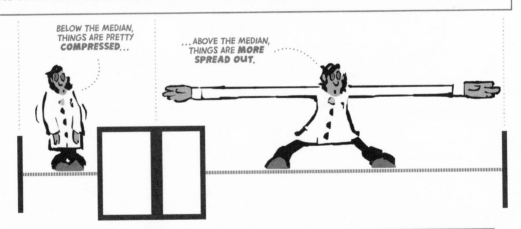

BELOW THE MEDIAN, THINGS ARE PRETTY **COMPRESSED**...

...ABOVE THE MEDIAN, THINGS ARE **MORE SPREAD OUT**.

...AND IS ESPECIALLY USEFUL FOR INVESTIGATING DATA THAT ARE **SKEWED**.

LOOKS LIKE OUR SUPERVILLAINS **SKEW OLD**, WATSON...

...BUT I WONDER HOW THINGS WILL LOOK IF **JIMMY THE GEEZER** EVER KICKS THE BUCKET?

...I'M AN **OUTLIER**.

THE MOST COMMON MEASURE OF SPREAD, HOWEVER, IS **STANDARD DEVIATION (SD).** *

IF WE THINK OF THE AVERAGE AS A **CENTRAL VALUE**...

...THE STANDARD DEVIATION IS BASED ON AN **AVERAGE DISTANCE FROM THAT VALUE.**

SD WORKS BEST WHEN THE PILE OF DATA IS FAIRLY **SYMMETRICAL.**

LIKE THIS ONE, WHICH DESCRIBES OUR **HEIGHT.**

STANDARD MEANS **TYPICAL**...

...DEVIATION MEANS **DIFFERENCE!**

Avg

SD

Supervillain Count

Height (in cm)   140   160   180   200

UNFORTUNATELY, **CALCULATING** STANDARD DEVIATION IS A BIT **TRICKY.**

WE TAKE THE **SQUARE ROOT** OF THE **AVERAGE SQUARED DIFFERENCE** FROM THE **AVERAGE VALUE!**

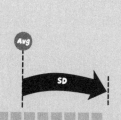

AAHHHH!

SO FOR NOW JUST REMEMBER THAT A **WIDER PILE OF DATA** HAS A **LARGER STANDARD DEVIATION.**

AND A LARGER STANDARD DEVIATION...

...MEANS **MORE VARIATION!**

Avg

SD

* SEE **PAGE 215** TO LEARN HOW TO CALCULATE IT.

IN THIS CHAPTER WE LEARNED ABOUT THE **FOUR IMPORTANT QUALITIES** TO LOOK FOR IN ANY **ONE SAMPLE**...

OKAY, HERE'S A RANDOM SAMPLE OF FISH!

GREAT, WHAT'S ITS **SAMPLE SIZE**...

...SHAPE...

...**LOCATION**...

...AND **SPREAD**?

...AND SOON WE'LL BE HUNTING FOR SOME OF **THESE SAME QUALITIES** IN A LARGER **POPULATION**.

THE POPULATION ALSO HAS **SHAPE**, **LOCATION**, AND **SPREAD**...

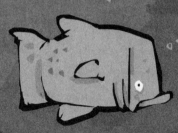

...YOU'LL JUST **NEVER KNOW THEM FOR CERTAIN!**

BUT FIRST, LET'S USE WHAT WE'VE LEARNED SO FAR TO **RESOLVE A CONTROVERSY!**

CHAPTER 5
MONSTER MISTAKES

MOST OF THE TIME WHEN WE GO OUT AND GATHER DATA,...

...IT'S BECAUSE WE'RE **INVESTIGATING AN IMPORTANT QUESTION ABOUT THE WORLD.**

WHEN DID THESE MOUNTAINS **RISE UP FROM THE SEA?**

HOW MANY BEHEADINGS HAPPENED DURING THE REIGN OF KING HENRY VIII?

WILL GIRLS DIG IT IF I **WEAR THESE PANTS?**

SOME QUESTIONS ARE FAIRLY **STRAIGHTFORWARD**...

...AND CAN BE TACKLED BY LOOKING AT ONLY **ONE SET OF SAMPLE DATA.**

HOW MANY PEOPLE IN THIS COUNTRY HAVE **DIABETES?**

LET'S EXAMINE **100 RANDOM CITIZENS** AND MAKE A GUESS.

DO VAMPIRES HAVE **BAD BREATH?**

LET'S EXAMINE **100 RANDOM VAMPIRES** AND MAKE A GUESS.

BUT OTHER QUESTIONS ARE MORE **COMPARATIVE**...

WHEN THEY **BITE DIABETIC PEOPLE**...

...DO **VAMPIRES GET BAD BREATH?**

...AND REQUIRE MORE **COMPLEX ANALYSIS.**

MORE **COMPLEX** STATISTICS PROBLEMS OFTEN INVOLVE **EXPLORING A RELATIONSHIP...**

ARE THESE TWO THINGS **RELATED?**

... BETWEEN **ONE VARIABLE**...

... AND **ANOTHER.**

DOES **DRINKING HIPPO SALIVA**...

...**CURE BALDNESS?**

DOES SLATHERING **ANTI-INFLAMMATORY STEROIDAL NOUGAT CREAM ON YOUR BEHIND**...

...AFFECT **YOUR INTELLIGENCE?**

DOES **WEARING A MAGNET ON YOUR HEAD**...

...MAKE PEOPLE **ATTRACTED TO YOU?**

MUCH OF THE TIME WE'RE EXPLORING HOW ONE VARIABLE **INFLUENCES** ANOTHER...

...BUT REMEMBER, WHEN WE USE STATISTICS WE CAN NEVER TOTALLY PROVE **ANY** OF OUR CONCLUSIONS.

DOES **EATING LOTS AND LOTS OF CARROTS**... ...CAUSE YOUR **SKIN TO TURN ORANGE?**

TO PROVE IT YOU'D HAVE TO FEED LOTS AND LOTS OF CARROTS TO **EVERYBODY** ON THE PLANET...

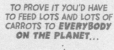

...SO LET'S FEED THEM TO **100 RANDOM SCHOOLCHILDREN** INSTEAD.

IN THIS CHAPTER WE'RE GOING TO EXPLORE A RELATIONSHIP **BETWEEN TWO DIFFERENT VARIABLES...** *

WOULD BEING A **WOMAN**... ...**MAKE ME FASTER?**

... TO **SETTLE AN ARGUMENT.**

IN THE OLD DAYS, ONLY **MALE VIKINGS** RODE DRAGONS.

WHOOOOOOO!

BUT RECENTLY, **FEMALE VIKINGS** HAVE STARTED RIDING...

YEEEEOOOOOOO!

... AND THEY'RE CONVINCED THAT **THEY'RE FASTER.**

EAT MY TOOTS, YOU **CHAUVINISTIC PIG!**

* SEE **PAGE 215** FOR A TECHNICAL DEFINITION.

IN ORDER TO TEST WHETHER **GENDER**...

...HAS A NOTICEABLE EFFECT ON **SPEED**...

OUR **FIRST VARIABLE**.

OUR **SECOND VARIABLE**.

...VIKING JUDGES **COLLECTED SOME DATA**.

THEY **RANDOMLY SELECTED 50 MALE** RIDERS...

...AND **50 FEMALE** RIDERS...

WE ROCK!

WE CHOSE THEM **RANDOMLY** SO AS NOT TO BIAS THE RESULTS...

...BY ACCIDENTALLY CHOOSING THE **FASTEST** RIDERS...

...OR THE **SLOWEST**.

YOU SUCK!

...AND TIMED RIDERS AS THEY ZIPPED AROUND **A KILOMETER-LONG COURSE**.

MAY THE **FASTEST** RIDERS WIN!

ALL TIMES ARE IN **SECONDS**.

HERE ARE THE TIMES POSTED BY 50 RANDOM **MALE RIDERS**...

...AND 50 RANDOM **FEMALE RIDERS**.

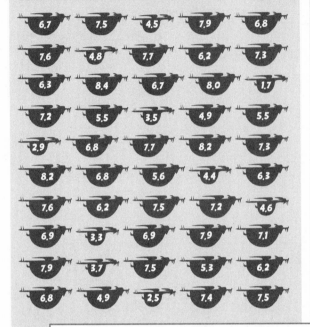

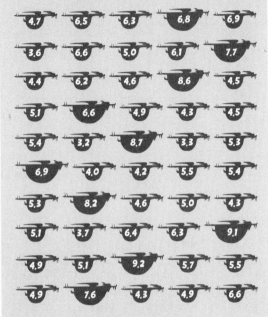

FROM THIS RAW DATA WE CAN EASILY **CALCULATE TWO SAMPLE AVERAGES**...

ADD UP THE 50 MALE TIMES...

...AND DIVIDE BY 50.

THE MALE RIDERS AVERAGED **6.3 SECONDS**.

ADD UP THE 50 FEMALE TIMES...

...AND DIVIDE BY 50.

THE FEMALE RIDERS AVERAGED **5.6 SECONDS**.

...AND **COMPARE THEM**.

ON AVERAGE, **THE FEMALE RIDERS ARE FASTER!**

NOTE THAT WE'LL LEARN LATER WHAT THIS SAMPLE DATA SUGGESTS ABOUT THE OVERALL POPULATIONS...

...BUT FOR NOW WE'RE JUST **FOCUSING ON THE SAMPLE DATA ITSELF.**

OUR SAMPLE AVERAGE IS BETTER THAN YOUR SAMPLE AVERAGE!

BUT SO FAR WE'VE LOOKED AT ONLY **ONE PART OF THE OVERALL PICTURE.**

BEWARE THE **HASTY AVERAGE!**

WE ALWAYS NEED TO LOOK AT SHAPE, LOCATION, **AND** SPREAD.

TO GET A MORE DETAILED SENSE OF THE DATA, WE SHOULD **ALWAYS DRAW PICTURES.**

GATHER ROUND AND LET'S **SEE WHAT THE NUMBERS REVEAL.**

SURE ENOUGH, A **DIFFERENT PICTURE EMERGES**...

BOTH GROUPS ARE **SKEWED!**

...WHEN WE COMPARE A **BOXPLOT** OF THE **MALE RIDERS' DATA**...

WE HAD THE **SLOWER TIMES** OVERALL...

...BUT WE HAD **MORE VARIATION** OVER THERE ON THE **FASTER SIDE!**

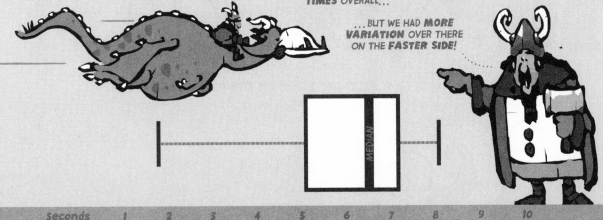

MEDIAN

Seconds 1 2 3 4 5 6 7 8 9 10

...WITH A **BOXPLOT** OF THE **FEMALE RIDERS' DATA**.

WE HAD THE **FASTER TIMES** OVERALL...

...BUT WE HAD **MORE VARIATION** OVER THERE ON THE **SLOWER SIDE!**

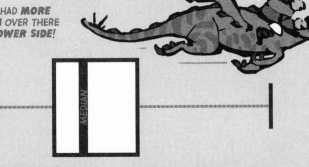

MEDIAN

Seconds 1 2 3 4 5 6 7 8 9 10

WHY WOULD **BOTH** GROUPS BE SKEWED...

...IN DIFFERENT DIRECTIONS?

I SUSPECT **HIDDEN FORCES** AT WORK!

AND THE **MYSTERY DEEPENS...**

BOTH GROUPS ALSO SEEM TO HAVE **DOUBLE HUMPS!**

... WHEN WE LOOK AT **HISTOGRAMS** OF THE **MALE DATA**...

THIS DATA HAS ONE **SMALL HUMP** HERE ON THE **FAST** SIDE...

...AND ONE **BIG HUMP** HERE ON THE **SLOW** SIDE.

WE CALL THIS TYPE OF DOUBLE-HUMPED SHAPE **BIMODAL**.

...AND THE **FEMALE DATA**.

CONVERSELY, THIS DATA HAS ONE **BIG HUMP** HERE ON THE **FAST** SIDE...

...AND ONE **SMALL HUMP** HERE ON THE **SLOW** SIDE...

...AND THERE MUST BE A **REASON WHY!**

THESE PICTURES SUGGEST THAT THE RELATIONSHIP BETWEEN OUR TWO VARIABLES **MAY NOT BE AS SIMPLE AS WE THOUGHT**.

IF **BEING A WOMAN**...

...MAKES YOU **GO FASTER**...

...WHY DO **BOTH** DATA SETS HAVE **SKEWED TAILS AND MYSTERIOUS DOUBLE HUMPS?**

REMEMBER, HOWEVER OUR DATA IS SHAPED, IT'S ALWAYS SHAPED THAT WAY **FOR A REASON**.

THE CHALLENGE NOW IS TO FIGURE OUT **WHY THE DATA LOOKS THE WAY IT DOES...**

SKEWED TAILS AND DOUBLE HUMPS?

I THINK WE'RE MISSING SOMETHING **MONSTROUS.**

... BY SEARCHING FOR **OTHER VARIABLES THAT MIGHT BE INFLUENCING IT.**

WHAT ELSE COULD BE **AFFECTING RIDER SPEED?**

COULD IT BE SOMETHING **ABOUT THE COURSE?**

COULD BE, BUT I DOUBT IT...

... SINCE THAT WAS **THE SAME FOR BOTH SETS OF RIDERS.**

COULD IT BE **HOW MUCH THE RIDERS WEIGH...**

... OR **WHAT THEY WEAR?**

COULD BE, BUT I DOUBT IT.

REMEMBER, WE CHOSE THEM **RANDOMLY.**

IT TURNS OUT THAT WHILE FOCUSING ON **GENDER** AND **SPEED...**

... WE'VE BEEN NEGLECTING TO THINK ABOUT **THE DRAGONS THEMSELVES!**

WE'RE A **THIRD VARIABLE!**

YOU CAN CHOOSE A BIG, MEAN, BURLY GRUNT...

...OR A NIMBLE, SLY, SPEEDY RUNT.

...AND MALE RIDERS TEND TO PREFER THE LARGER, **SLOWER** DRAGONS...

WE DON'T RIDE NO **WIMPY DRAGONS!**

...WHILE FEMALE RIDERS PREFER THE SMALLER, **FASTER** DRAGONS!

WELL, DUH... ...THEY'RE **FASTER.**

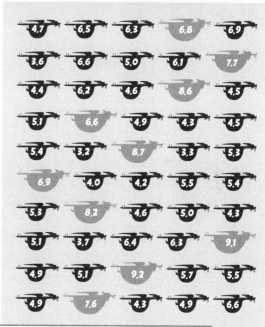

SO IT'S NO WONDER THE FEMALE RIDERS **SEEMED FASTER OVERALL.**

80% OF US CHOSE **SLOWER DRAGONS!**

80% OF US CHOSE **FASTER DRAGONS!**

WHEN WE ACCOUNT FOR **DRAGON PREFERENCE**...

I LIKE **BIG DRAGONS**, YOU GOT A PROBLEM WITH THAT?!

... BY CALCULATING AVERAGE TIMES BY **BOTH RIDER GENDER AND DRAGON TYPE**...

... THE RESULTS ARE **SURPRISING**.

WE SIFTED THROUGH THE RAW DATA AND FOUND THE AVERAGES TO GET THESE VALUES.

THE MALE RIDERS TENDED TO BE FASTER ON **BOTH KINDS OF DRAGONS!**

	SMALL DRAGONS	BIG DRAGONS
MALE RIDERS	3.6	6.9
FEMALE RIDERS	5.1	7.9

IT TURNS OUT THAT OUR **FIRST CONCLUSION**...

... WAS NOT ONLY **MISLEADING**...

WE THOUGHT **FEMALE RIDERS** WERE FASTER...

... BUT WE WERE JUST **CHOOSING TO RIDE FASTER DRAGONS!**

... IT WAS **DEAD WRONG!**

YOU MAY BE FASTER **OVERALL**...

... BUT WHEN WE TAKE **DRAGON CHOICE** INTO ACCOUNT...

... **THEY'RE** FASTER!

WHILE WE WERE BUSY LOOKING FOR A RELATIONSHIP BETWEEN **TWO VARIABLES**...

...A **HIDDEN, LURKING VARIABLE** WAS SNEAKING AROUND...

DOES **RIDER GENDER**...

...**AFFECT SPEED**?

...**WREAKING HAVOC** WITH OUR CONCLUSIONS.

MY ADVICE, **DON'T IGNORE THE DRAGONS**.

UNFORTUNATELY, LURKING VARIABLES CAN **BEDEVIL ALL KINDS OF STATISTICAL ANALYSIS**...

THE UNIVERSE IS **FULL OF VARIABLES**!

SOME WE ALREADY KNOW ABOUT...

...AND **SOME WE DON'T**.

...AND PART OF A STATISTICIAN'S JOB IS TO **DIG AROUND FOR THEM**.

THE **MORAL** OF THIS STORY...

BEWARE **LURKING VARIABLES!**

...IS THAT **WHENEVER WE THINK** WE SEE A RELATIONSHIP BETWEEN **ANY TWO VARIABLES**...

... THERE MIGHT BE SOME OTHER VARIABLE **INFLUENCING OUR CONCLUSIONS**...

EATING ONLY CABBAGES...

...EXTENDS YOUR LIFE SPAN!

...UNLESS PEOPLE WHO EAT ONLY CABBAGES **TEND TO EXERCISE MORE.**

EARNING FEWER DOUBLOONS...

...MAKES **PIRATES** ANGRY!

...UNLESS IT'S THE **WRETCHED FOOD** THAT MAKES US ANGRY!

STORKS...

...DELIVER BABIES.

...UNLESS THEY ARRIVE BY **OTHER MEANS.**

... AND IF WE DON'T FIND IT, WE RISK BELIEVING THINGS THAT **AREN'T TRUE!**

CHAPTER 6
FROM SAMPLES TO POPULATIONS

THUS FAR, WE'VE BEEN TALKING ALMOST EXCLUSIVELY ABOUT **SAMPLES**.

HERE ARE 50 RANDOM FISH, SORTED BY WEIGHT!

BUT REMEMBER, OUR **ULTIMATE TASK** IS TO USE **SAMPLES**...

...WHICH **WE CAN LOOK AT**!

... TO MAKE **CONFIDENT STATEMENTS** ABOUT **ENTIRE POPULATIONS**.

...WHICH WE'LL **NEVER** BE ABLE TO LOOK AT!

THIS POSES A **PROBLEM:**

HOW CAN WE
BE CONFIDENT ABOUT
POPULATIONS...

...WHEN WE'LL
NEVER BE ABLE
TO LOOK AT THEM?

IT'S **MURKY**
DOWN THERE.

IN **PART TWO OF THIS BOOK** WE'RE GOING TO
TACKLE THIS PROBLEM HEAD-ON...

WE'RE GOING TO
LEARN **STATISTICAL
INFERENCE!**

...BUT BEFORE WE BEGIN, LET'S CLARIFY SOME OF THE **KEY TERMS** WE'LL BE USING.

WE'VE LEARNED THAT WHEN WE PILE UP OUR SAMPLE DATA...

... WE CALL THE RESULT A **SAMPLE HISTOGRAM.**

HERE ARE **THE 50 RANDOM FISH WE ACTUALLY CAUGHT.**

THIS SHOWS **SOME OF US.**

BUT IF WE COULD SOMEHOW PILE UP AN **ENTIRE POPULATION...**

... WE'D CALL THE RESULT A **POPULATION DISTRIBUTION.***

HERE ARE **ALL THE FISH IN THE WHOLE LAKE!**

THIS SHOWS **ALL OF US.**

REMEMBER, IN REALITY, YOU **NEVER ACTUALLY GET TO SEE** AN ENTIRE POPULATION DISTRIBUTION...

...IF YOU COULD, YOU **WOULDN'T NEED STATISTICS.**

* SEE PAGE 216.

SIMILARLY, WE'VE LEARNED THAT **SAMPLE HISTOGRAMS HAVE CERTAIN IMPORTANT QUALITIES...**

THE PILE OF **FISH WE CAUGHT...**

...HAS A **SHAPE...**

...**LOCATION...**

AVG =3.7

...AND **SPREAD...**

SD=1.9

...AND **WE KNOW THEM!**

...AND IT TURNS OUT THAT **POPULATION DISTRIBUTIONS ALSO HAVE THESE QUALITIES.**

THE **ENTIRE POPULATION** OF FISH IN THE LAKE...

?

...ALSO HAS A **SHAPE, LOCATION,** AND **SPREAD...**

AVG =?

SD=?

...BUT WE'LL **NEVER KNOW THEM FOR CERTAIN.**

IN ORDER TO DISTINGUISH BETWEEN THEM, WE REFER TO QUALITIES IN **SAMPLES** AS **"STATISTICS"...**

FOR EXAMPLE, OUR **SAMPLE AVERAGE IS A STATISTIC...**

...AND SO IS OUR **SAMPLE STANDARD DEVIATION.**

...AND TO QUALITIES IN **POPULATIONS** AS **"PARAMETERS."*

FOR EXAMPLE, OUR **OVERALL POPULATION AVERAGE IS A PARAMETER...**

...AND SO IS OUR **POPULATION STANDARD DEVIATION.**

IN OTHER WORDS, THE ONLY REASON WE GO OUT AND **GATHER STATISTICS**...

...IS BECAUSE WE'RE **CURIOUS ABOUT PARAMETERS**.

WE KNOW THAT THESE RANDOM SAMPLE FISH HAVE AN AVERAGE WEIGHT OF **3.7 POUNDS**...

...BUT WHAT WE **REALLY CARE ABOUT** IS WHETHER THAT'S TRUE OF **ALL THE FISH IN THE LAKE**.

STATISTICS ARE THE THINGS WE **ACTUALLY CALCULATE** AND THEREFORE KNOW WITH **CERTAINTY**...

...AND **PARAMETERS** ARE THE THINGS WE REALLY **WANT TO KNOW** BUT CAN **ONLY MAKE GUESSES ABOUT**.

STATISTICS ARE THE NUMBERS WE'RE LOOKING **AT**.

PARAMETERS ARE THE NUMBERS WE'RE LOOKING **FOR**.

LUCKILY, ALTHOUGH WE'LL **NEVER** BE ABLE TO **LOOK AT PARAMETERS DIRECTLY**...

...WE CAN **USE STATISTICS** TO **HUNT FOR THEM**.

GRAB THE STATISTICS...

...WE'RE GOING ON A **MISSION!**

IN PRACTICE, STATISTICIANS HUNT FOR **ALL KINDS OF DIFFERENT PARAMETERS.**

STANDARD DEVIATIONS.

PROPORTIONS.

MEDIANS, YOU NAME IT!

BUT WE'RE GOING TO FOCUS ON **ONE IN PARTICULAR.**

WE'RE GOING TO LEARN TO USE STATISTICS WE FIND IN **ONE RANDOM SAMPLE...**

...TO HUNT FOR THE **AVERAGE IN THE POPULATION IT COMES FROM.**

SAMPLE SIZE?

CHECK.

SAMPLE AVERAGE?

CHECK.

SAMPLE STANDARD DEVIATION?

CHECK.

OKAY, WE'RE **READY TO HUNT!**

IT'S **GOT TO BE AROUND HERE SOMEWHERE!**

AS WE KNOW, WE'LL **NEVER** BE ABLE TO USE **STATISTICS**...

...TO ACHIEVE **CERTAINTY ABOUT PARAMETERS**.

I CAN EXAMINE THESE **50 RANDOM ALIENS** I JUST GATHERED FROM THAT GALAXY.

BUT THERE ARE A **GAZILLION MORE** ALIENS IN THIS GALAXY THAT SHE'LL NEVER BE ABLE TO EXAMINE.

FORTUNATELY, STATISTICIANS HAVE DISCOVERED A WAY TO **BRIDGE THE DISTANCE BETWEEN THEM**...

WHOA! WE CAN USE **THIS SHAPE** LIKE A **LENS**...

...TO **BRING POPULATIONS INTO FOCUS!**

...AND WE'RE GOING TO SPEND THE **NEXT CHAPTER** LEARNING ABOUT IT!

PART TWO
HUNTING PARAMETERS

CHAPTER 7
THE CENTRAL LIMIT THEOREM

THIS CHAPTER IS ABOUT THE **GREAT DISCOVERY**...

...THAT MAKES **EVERYTHING IN THE REST OF THIS BOOK POSSIBLE**...

...AND IT HAS TO DO WITH **AVERAGES**.

LET'S IMAGINE THAT WE WANT TO KNOW THE **AVERAGE VALUE IN A CERTAIN POPULATION.**

HOW MUCH **SODA** DOES EACH AMERICAN **DRINK PER DAY?**

SLURP.

BURP!

THEN LET'S IMAGINE WE GO OUT AND GATHER **A WHOLE BUNCH OF SEPARATE RANDOM SAMPLES FROM THAT POPULATION.**

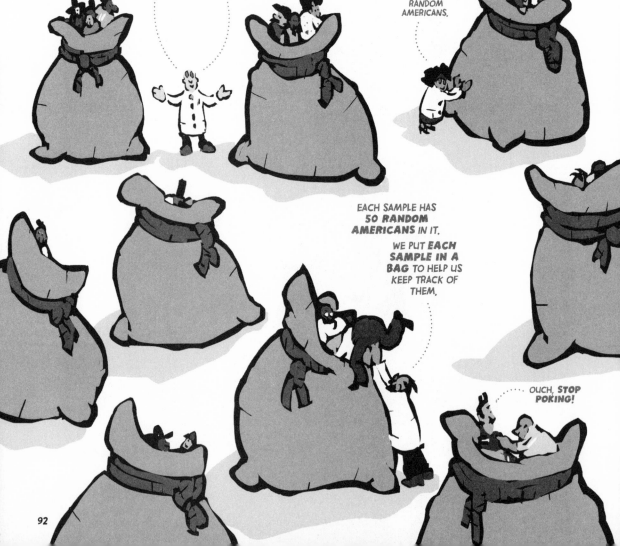

HERE ARE **50 RANDOM AMERICANS.**

HERE ARE 50 **OTHER** RANDOM AMERICANS.

HERE ARE 50 **OTHER** RANDOM AMERICANS.

EACH SAMPLE HAS **50 RANDOM AMERICANS** IN IT.

WE PUT **EACH SAMPLE IN A BAG** TO HELP US KEEP TRACK OF THEM.

OUCH, **STOP POKING!**

IT TURNS OUT THAT IF WE CALCULATE THE **AVERAGE VALUE** IN EACH OF OUR **RANDOM SAMPLES**...

FOR EXAMPLE, THE AVERAGE IN OUR SAMPLE IS **17.2 OUNCES.**

IN OUR SAMPLE IT'S **12.9 OUNCES.**

HERE, IT'S **18.4 OUNCES.**

6.3

16.1

15.3

17.2

12.9

18.4

6.3

... THEN ORDER THEM AND **PILE THEM UP**...

WE BUILD A HISTOGRAM **WITH THE AVERAGES.**

12.9

18.4

20.3

OOF!

Average Daily Soda Intake   10   15   20   25

... **THE PILE OF AVERAGES** WILL EVENTUALLY START TO **CLUMP TOGETHER!**

WE CAN EXPECT TO SEE **SOME EXTREME AVERAGE VALUES** LIKE THIS ONE.

BUT **MOST OF THE AVERAGES CLUMP AROUND HERE.**

BETWEEN 15 AND 20 OUNCES PER DAY.

HMMMM.

16.8

17.4   19.8

12.7   15.3   19.1

6.3

12.9   17.2   18.4   20.3

22.1

Average Daily Soda Intake   10   15   20   25

AND THAT'S **NOT ALL.**

93

IT TURNS OUT THAT AS YOU PILE UP **MORE AND MORE SAMPLE AVERAGES...**

BRING MORE!

WE WANT **GAZILLIONS!**

... THE WHOLE PILE WILL TEND TO GET **MORE AND MORE NORMAL-SHAPED.**

REMEMBER, EACH BAG IS A SEPARATE SAMPLE...

... AND WE'RE SORTING THEM BY AVERAGE VALUE PER BAG.

THIS IS A **GREAT DISCOVERY!**

17.3
15.9
16.9
14.4 15.1
17.6
14.1 16.8 17.7
13.3 17.4 19.8
12.2 12.7 15.3 19.1 22.1
17.9
6.3 10.8 12.9 17.2 18.4 20.3 22.8

Average Daily Soda Intake   10   15   20   25

THIS **NORMAL SHAPE** HAS VERY **PRECISE MATHEMATICAL FEATURES.** *

BUT FOR NOW JUST KEEP IN MIND THAT IT'S SHAPED LIKE A **SYMMETRICAL BELL.**

IN FACT, IT LOOKS EXACTLY LIKE **THIS!**

$$h_{\mu,\sigma}(x) = \frac{1}{\sigma\sqrt{2\pi}} \exp\left\{-\frac{1}{2\sigma^2}(x-\mu)^2\right\}$$

94

* SEE **PAGE 217** FOR SOME TECHNICAL DETAILS.

IT WORKS FOR RANDOM SAMPLE AVERAGES FROM **ANY POPULATION.**

RANDOM **DRAGON** SCALE SAMPLES...

...SORTED BY THEIR **AVERAGE WEIGHT.**

RANDOM **LIZARD** LEG SAMPLES...

...SORTED BY THEIR **AVERAGE LENGTH.**

AND IT **DOESN'T MATTER** HOW THE **POPULATION ITSELF** IS SHAPED.

IT COULD BE SHAPED LIKE **THIS**...

...OR **THIS!**

FLAT, SKEWED, NORMAL, ABNORMAL, WHATEVER!

IN THE LONG RUN, THE MORE **AVERAGES** YOU PILE UP, **THE MORE NORMAL-SHAPED THE PILE TENDS TO BECOME.**

BELL CURVED AND SYMMETRICAL!

AND TRAILING OFF DELICATELY AT BOTH ENDS, JUST LIKE THIS.

THIS IS THE MOST BEAUTIFUL SHAPE IN ALL OF STATISTICS!

Frequency

Sample Averages

TECHNICALLY, A HUGE PILE LIKE THIS IS A TYPE OF **SAMPLING DISTRIBUTION.***

IT'S HOW **SAMPLE STATISTICS** WOULD BE **DISTRIBUTED**...

...IF WE GATHERED A GAZILLION OF THEM.

* SEE **PAGE 217** FOR A DEFINITION.

PLUS, THERE'S **AN ADDED BONUS:**

SHE'S THE MOST **BEAUTIFUL SHAPE** IN ALL STATISTICS...

... **AND** SHE LIKES HEAVY METAL MUSIC!?

IT TURNS OUT THAT THE **CENTER VALUE** IN A **GIANT PILE OF AVERAGES...**

THAT'S **THE AVERAGE** OF ALL THE AVERAGES!

BRING MORE SAMPLES!

THIS ONLY WORKS WHEN WE HAVE **LOTS** AND **LOTS** AND **LOTS** OF THEM!

... IS EQUAL TO THE **CENTER VALUE OF THE POPULATION IT CAME FROM.**

THE AVERAGE OF ALL THE AVERAGES...

... EQUALS **THE POPULATION AVERAGE.**

THE POPULATION MIGHT BE SHAPED LIKE **THIS**...

... WE'LL **NEVER KNOW FOR CERTAIN.**

FOR EXAMPLE, IF THIS GIANT PILE OF SAMPLE AVERAGES SORTED BY DAILY SODA INTAKE IS CENTERED AT **17 OUNCES PER DAY**...

ADSI 16 **17** 18 19

...THE OVERALL POPULATION WILL BE **CENTERED AT THAT SAME VALUE!**

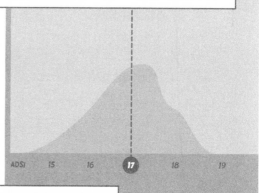

ADSI 15 16 **17** 18 19

THIS WORKS BECAUSE A GIANT PILE OF AVERAGES IS **GUARANTEED TO BE SYMMETRICAL.**

NORMAL DISTRIBUTIONS **ARE ALWAYS SYMMETRICAL.**

IN THE LONG RUN, FOR **EACH** SAMPLE AVERAGE WE GRAB WITH A VALUE **BELOW THE POPULATION AVERAGE**...

...WE'RE GUARANTEED TO EVENTUALLY GRAB **ANOTHER** SAMPLE AVERAGE WITH A VALUE **ABOVE THE POPULATION AVERAGE.**

THESE 50 RANDOM AMERICANS **DON'T DRINK MUCH SODA.**

THESE 50 RANDOM AMERICANS **DRINK LOADS OF SODA.**

AND THERE'S A **SECOND ADDED BONUS:**

I HAVE **DIED** AND **GONE TO HEAVEN!**

IT TURNS OUT THAT **A GIANT PILE OF AVERAGES...**

REMEMBER, THIS PILE HAS A **GAZILLION** SAMPLES IN IT.

...WILL ALSO TEND TO BE **NARROWER** THAN THE **POPULATION IT CAME FROM.**

IN OTHER WORDS, THE PILE OF AVERAGES HAS A **SMALLER SPREAD...**

...WHICH MEANS **LESS VARIATION!**

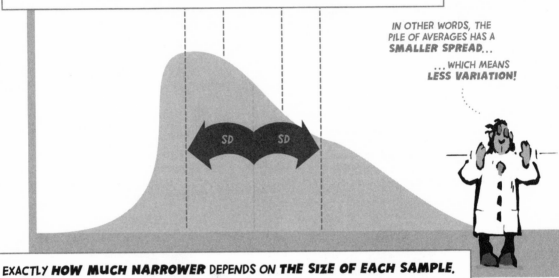

SD   SD

EXACTLY **HOW MUCH NARROWER** DEPENDS ON **THE SIZE OF EACH SAMPLE.**

AS WE INCREASE THE SAMPLE SIZE, THE GIANT PILE WILL LOOK LESS LIKE **THIS...**

...AND MORE LIKE **THIS.**

TALL AND SLENDER.

SHORT AND WIDE.

NOTE THAT **BOTH OF THESE PILES ARE NORMAL-SHAPED.**

BUT THE SLENDER ONE HAS A **SMALLER STANDARD DEVIATION.**

THERE'S AN INTUITIVE WAY TO THINK ABOUT WHY A **LARGER SAMPLE SIZE** RESULTS IN A **NARROWER PILE OF AVERAGES.**

HOP IN.

IF EACH SAMPLE HAS ONLY **ONE AMERICAN** IN IT...

...THEN THE SPREAD OF THE PILE OF AVERAGES **WILL BE EXACTLY THE SAME AS THE SPREAD OF THE WHOLE POPULATION.**

ONE SAMPLE PER BAG.

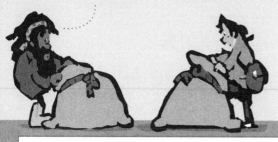

THE VARIATION **BETWEEN BAGS...**

...WILL **EQUAL** THE VARIATION **BETWEEN INDIVIDUALS IN THE OVERALL POPULATION!**

BUT IF EACH SAMPLE HAS **ALL AMERICANS** IN THE WHOLE POPULATION JAMMED INTO IT...

...THEN THE SPREAD OF THE PILE OF AVERAGES WILL BE **ZERO.**

ONE SAMPLE PER BAG.

THERE WILL BE **NO VARIATION** BETWEEN BAGS!

DUH!

IN ANY CASE, THE MATHEMATICAL RELATIONSHIP IS **VERY PRECISE.**

THE STANDARD DEVIATION OF THE ENORMOUS PILE...

...EQUALS THE **POPULATION STANDARD DEVIATION...**

...DIVIDED BY THE **SQUARE ROOT OF THE SAMPLE SIZE!**

AAHHHH!

HERE'S A SUMMARY OF THE **GREAT DISCOVERY:**

IN THE LONG RUN, GIANT PILES OF RANDOM SAMPLE AVERAGES TEND TO BE NORMAL-SHAPED!

REMEMBER, EACH SAMPLE IS THE **SAME SIZE.**

THEY ALL COME FROM THE **SAME POPULATION.**

AND THERE'S A **GAZILLION** OF THEM!

YOU ARE THE **SAMPLING DISTRIBUTION OF MY DREAMS.**

JUST LOOK AT THOSE **BEAUTIFUL CURVES.**

Number of Baskets (75 random flowers per basket)

Flower Baskets Sorted by Average Length

THEY'RE **CENTERED AROUND THE POPULATION AVERAGE...**

... BUT **NARROWER THAN THE POPULATION.**

AND IT DOESN'T MATTER WHAT THE **SAMPLES...**

...OR THE **POPULATION...**

...ARE **SHAPED LIKE!**

TECHNICALLY,
WE CALL THIS DISCOVERY
**THE CENTRAL LIMIT
THEOREM (CLT).***

* SEE **PAGES 217–218** TO
READ ABOUT THE CLT AND
THESE CONDITIONS IN
GREATER DETAIL.

I WISH IT HAD A **MORE POETIC NAME.**

OVER THE YEARS, STATISTICIANS HAVE THROWN TOGETHER SOME MATH THAT
PROVES EXACTLY **WHY THE CLT WORKS**.

DOUBLE, DOUBLE
TOIL AND TROUBLE;
**RANDOM AVERAGE
SAMPLES BUBBLE.**

EYE OF
NEWT!

TOE OF
FROG!

SCALE OF
DRAGON!

TONGUE OF
DOG!

BUT THEY'VE ALSO DISCOVERED THAT THERE ARE **A FEW CONDITIONS**.

IT **ONLY WORKS** IF EACH SAMPLE IS TAKEN **RANDOMLY**...

THERE CAN BE NOTHING
EXCEPT CHANCE THAT MAKES
ANY ONE SAMPLE DIFFERENT
FROM ANY OTHER SAMPLE.

... **AND** IF EACH SAMPLE IS **LARGE ENOUGH**.

A SAMPLE SIZE
OF **30 OR MORE** IS
USUALLY SUFFICIENT...

...BUT IT DEPENDS
ON SOME OTHER
COMPLICATED MATH.

* CHECK OUT **PAGE 217** TO LEARN HOW TO SAY ALL THIS WITH MATH SYMBOLS.

HERE'S WHAT THE CLT MEANS IN PRECISE **MATHEMATICAL TERMS:** *

WE CAN EXPECT GIANT PILES OF SAMPLE AVERAGES TO BE **NORMAL...** 1

...AND **CENTERED AT THE POPULATION AVERAGE...**

...WITH A **STANDARD DEVIATION** EQUAL TO...

...**THE POPULATION STANDARD DEVIATION** DIVIDED BY THE **SQUARE ROOT OF THE SAMPLE SIZE.** 2

WHEW!

1. BUT ONLY IF THE SAMPLES ARE TAKEN **RANDOMLY** AND THE **SAMPLE SIZE IS LARGE ENOUGH** (GENERALLY LARGER THAN 30 OR SO).

2. IF YOU LIKE MATH, NOTE THAT THIS IS THE REASON THE WHOLE PILE WILL BE **NARROWER** IF THE SAMPLE SIZE IS **LARGER**.

KEEP THIS **BLUEPRINT,** BECAUSE WE'LL BE **USING IT LATER.**

IN THE LONG RUN, **RANDOM SAMPLE AVERAGES** TEND TO **CLUSTER AROUND THE POPULATION AVERAGE...**

...IN THIS BEAUTIFUL SHAPE!

IN THE NEXT SEVERAL CHAPTERS, WE'RE GOING TO **LEARN WHY IT MATTERS.**

IT GIVES US SOMETHING WE CAN BE CONFIDENT ABOUT!

CHAPTER 8
PROBABILITIES

NOW WE
CAN START
HUNTING!

IN THE LAST CHAPTER, WE LEARNED THAT **GIANT PILES OF SAMPLE AVERAGES...**

... TEND TO HAVE A **NORMAL SHAPE**.

WE CAN BE **CONFIDENT** ABOUT THIS...

... IF THE SAMPLES ARE TAKEN **RANDOMLY**...

...**AND** THE SAMPLE SIZE IS **LARGE ENOUGH!**

Number of Jars (500 random olives per jar)

Olive Jars sorted by average weight per jar

NOW WE'RE GOING TO LEARN **WHY THAT MATTERS...**

WHAT MAKES **THAT SHAPE** SO **SPECIAL?**

...BY EXAMINING A **GIANT PILE OF SAMPLE AVERAGES...**

...INSIDE **CRAZY BILLY'S BAIT BARN**.

HI, I'M **BILLY**.

CAREFUL... ...HE'S **CRAZY!**

Crazy Billy's **BAIT BARN**

CRAZY BILLY IS CALLED CRAZY BILLY BECAUSE HE SPENDS AN **INSANE** AMOUNT OF TIME **CATCHING RANDOM WORM SAMPLES...**

30 WORMS PER SAMPLE,

I GRAB THEM **TOTALLY RANDOMLY...**

...FROM THE OVERALL POPULATION IN THE SWAMP,

...PUTTING EACH SAMPLE INTO A CAN...

BEFORE I SEAL EACH CAN I MEASURE THE WORMS...

...AND **CALCULATE THE AVERAGE WORM LENGTH PER CAN,**

...AND VERY CAREFULLY **PILING UP GAZILLIONS OF THOSE CANS,** EACH ACCORDING TO ITS AVERAGE VALUE...

...INSIDE HIS ENORMOUS BAIT BARN...

THIS CAN HAS AN AVERAGE LENGTH OF **4.75 INCHES,** SO I'LL PUT IT **EXACTLY HERE,**

Number of Cans

Average Length per Can (inches)   3.5   4   4.5   5

...OR SO HE CLAIMS,

YOU HAVE AN **ACTUAL SAMPLING DISTRIBUTION** IN THERE?

YUP, IT'S ALL BEHIND THAT **DOOR.**

IN THIS CHAPTER WE'RE GOING TO FIND OUT **WHAT WE CAN LEARN FROM CRAZY BILLY'S HUGE PILE.**

IT'S **NORMAL!**

IT'S CENTERED AT **4 INCHES.**

ITS STANDARD DEVIATION IS **0.25 INCHES.**

SO?

MORE SPECIFICALLY, WE'RE GOING TO FOCUS ON **THIS QUESTION:**

IF WE HAVE ACCESS **ONLY** TO **WHAT'S INSIDE THE BARN...**

...WHAT CAN WE SAY ABOUT THE **OVERALL POPULATION OF WORMS IN THE SWAMP?**

WHAT DO BILLY'S CANS OF WORMS...

...TELL US ABOUT ALL THE OTHER WORMS STILL ROAMING FREE?

HERE'S THE SAME QUESTION IN **TECHNICAL TERMS:**

IF WE HAVE ACCESS **ONLY** TO A **SAMPLING DISTRIBUTION** MADE UP OF AVERAGES...

...WHAT CAN WE SAY ABOUT THE OVERALL **POPULATION?**

THE FIRST **COOL THING** WE COULD FIGURE OUT IF WE COULD PEEK INTO BILLY'S BARN...

...IS THE OVERALL **POPULATION AVERAGE**.

REMEMBER, IN THE LONG RUN, SAMPLE AVERAGES TEND TO CLUSTER **AROUND THE POPULATION AVERAGE**.

SO THE ENTIRE SWAMP POPULATION AVERAGE IS **SMACK-DAB IN THE MIDDLE OF MY HUGE PILE!**

RIGHT HERE AT **4 INCHES**.

Average Length per Can (inches)   3.5   4   4.5   5

IN OTHER WORDS, IF WE WERE OUT **HUNTING FOR THE POPULATION AVERAGE IN THE SWAMP**...

...WE COULD **LOOK INSIDE THE BARN** TO FIND IT!

WHAT'S THE **AVERAGE LENGTH OF THE WORMS** IN THIS SWAMP?

NO NEED TO **SOIL YOUR SUIT** DIGGING IN THE MUCK.

BUT **THAT'S NOT ALL**...

THE **OTHER COOL THING** WE COULD DO IF WE COULD PEEK INTO BILLY'S BARN...

AND THIS IS THE **REALLY IMPORTANT COOL THING!**

...IS **CALCULATE PROBABILITIES** ABOUT THE OVERALL POPULATION!

WHAT'S A **PROBABILITY?**

IT'S A FANCY WORD FOR **LIKELIHOOD** OR **CHANCE.**

HERE'S HOW IT WORKS:

IF WE WERE ABLE TO **COUNT** ALL THE CANS IN BILLY'S **HUGE PILE**...

...AND FIND THAT **50% OF THEM** HAVE AN AVERAGE VALUE **WITHIN THIS RANGE**...

THAT'S ALL THE CANS IN THIS **DARKER SHADED PORTION**...

...WITH AVERAGES BETWEEN 3.75 AND 4.25 INCHES LONG.

REMEMBER, EACH CAN HAS 30 **RANDOM WORMS** IN IT.

Average Length per Can (inches)   3   3.5   4   4.5   5

...IT WOULD MEAN THAT IF WE **RANDOMLY ASSEMBLED ONE CAN FROM THE POPULATION**...

...THERE WOULD BE A **50% PROBABILITY** IT WOULD HAVE AN AVERAGE VALUE **WITHIN THAT SAME RANGE!**

30 **RANDOM WORMS**, COMING RIGHT UP!

WHICH MEANS THERE'S A **50% CHANCE** IT'LL HAVE AN AVERAGE WORM LENGTH BETWEEN **3.75 AND 4.25 INCHES LONG!**

ALTERNATIVELY, IF WE COUNTED ALL THE CANS AND DISCOVERED **THIS ABOUT BILLY'S PILE:**

IT WOULD MEAN **THIS ABOUT THE OVERALL POPULATION:**

95% OF ALL THE CANS...

...HAVE AN AVERAGE VALUE **BETWEEN 3.5 AND 4.5 INCHES!**

THERE'S A **95%** PROBABILITY THAT THE NEXT CAN WE RANDOMLY ASSEMBLE FROM THE SWAMP...

...WILL HAVE AN AVERAGE VALUE **BETWEEN 3.5 AND 4.5 INCHES!**

AND IF **THIS** WERE TRUE ABOUT THE PILE:

THIS WOULD BE TRUE ABOUT THE POPULATION:

5% OF ALL THE CANS...

...HAVE AN AVERAGE VALUE THAT'S **LESS THAN 3.5 OR GREATER THAN 4.5 INCHES!**

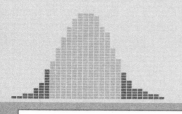

THERE'S A **5%** PROBABILITY THAT THE NEXT CAN WE RANDOMLY ASSEMBLE FROM THE SWAMP...

...WILL HAVE AN AVERAGE VALUE THAT'S **LESS THAN 3.5 OR GREATER THAN 4.5 INCHES!**

IN OTHER WORDS, BY **PEEKING INSIDE THE BARN...**

... WE CAN **MEASURE OUR CHANCES** OF RANDOMLY GATHERING RANGES OF AVERAGES **FROM THE SWAMP!**

MY PILE IS LIKE A **CRYSTAL BALL!**

I CAN USE IT TO TELL YOU WHAT SORT OF AVERAGE YOU'LL **PROBABLY** GRAB NEXT!

Crazy Billy's BAIT BARN

THERE ARE A FEW THINGS TO KEEP IN MIND WHEN WE **CALCULATE PROBABILITIES.** *

OMMMM.

* SEE **PAGE 218** FOR MORE TECHNICAL DETAILS.

FIRST, PROBABILITIES APPLY ONLY TO THE **LONG RUN**...

...SO THEY'LL **NEVER GIVE US CERTAINTY** ABOUT THE SHORT RUN.

FOR EXAMPLE, IF THERE'S A **95% PROBABILITY** THAT THE NEXT CAN WE RANDOMLY ASSEMBLE FROM THE SWAMP...

...WILL HAVE AN AVERAGE VALUE **INSIDE THIS RANGE**...

...IT DOESN'T MEAN OUR **NEXT CAN** WILL **NECESSARILY** HAVE AN AVERAGE INSIDE THAT RANGE!

IT JUST MEANS IT'S **VERY LIKELY** BECAUSE IN THE LONG RUN 19 OUT OF 20 DO!

SECOND, EVERY PROBABILITY HAS A **FLIP SIDE**...

FOR EXAMPLE, IF THERE'S A **50% PROBABILITY** THAT THE NEXT CAN WE RANDOMLY ASSEMBLE FROM THE SWAMP...

...WILL HAVE AN AVERAGE VALUE **INSIDE THIS RANGE**...

...THERE'S **ALSO A 50% PROBABILITY** THAT THE NEXT CAN WE RANDOMLY ASSEMBLE FROM THE SWAMP...

...WILL HAVE AN AVERAGE VALUE **OUTSIDE THAT RANGE!**

...BECAUSE PROBABILITIES ALWAYS **ADD UP TO 100%**.

IF THERE'S A **50% CHANCE** ONE THING WILL HAPPEN...

...THERE'S ALWAYS GONNA BE A **50% CHANCE SOMETHING ELSE** WILL HAPPEN.

IF THERE'S A **95% CHANCE** ONE THING WILL HAPPEN...

...THERE'S ALWAYS GONNA BE A **5% CHANCE SOMETHING ELSE** WILL HAPPEN.

FINALLY, BY DEFINITION, WE CAN CALCULATE PROBABILITIES ONLY **ABOUT RANDOM EVENTS...**

BY DEFINITION, A **PROBABILITY** IS A NUMBER THAT **QUANTIFIES THE LONG-TERM LIKELIHOOD** THAT A **CERTAIN RANDOM EVENT** WILL OCCUR.

...AND THAT'S WHY WE ALWAYS **GATHER STATISTICS RANDOMLY.**

IF I DIDN'T GATHER MY WORMS **RANDOMLY...**

...THE PILE IN MY BARN WOULDN'T MEAN DIDDLY.

MORE GENERALLY, WE CAN CALCULATE PROBABILITIES ABOUT **OTHER RANDOM EVENTS,** LIKE **COIN FLIPS...**

THE PROBABILITY OF **FLIPPING A COIN** AND GETTING **HEADS...**

...IS **50%...**

...BECAUSE **IN THE LONG RUN,** WE CAN EXPECT 50% OF ALL COIN FLIPS TO LAND ON HEADS.

...AND **ROLLS OF THE DICE.**

THE PROBABILITY OF **ROLLING A DIE** AND GETTING **A SIX...**

...IS **1/6...**

...BECAUSE **IN THE LONG RUN,** WE CAN EXPECT 1/6 OF ALL DIE ROLLS TO LAND ON SIX.

BUT LET'S RETURN TO **RANDOM WORM HUNTING...**

BLINDFOLDS ON!

...BECAUSE WE HAVE **ONE MORE REALLY IMPORTANT THING** TO LEARN ABOUT BILLY'S BARN!

IT TURNS OUT THAT WE DON'T HAVE TO ACTUALLY **COUNT** ALL THE INDIVIDUAL CANS IN BILLY'S BARN...

A GAZILLION AND **ONE**...
A GAZILLION AND **TWO**...
A GAZILLION AND **THREE**...

... TO CALCULATE PROBABILITIES.

95% OF ALL MY CANS HAVE AN AVERAGE BETWEEN 3.5 AND 4.5 INCHES!

IT TURNS OUT THAT BECAUSE WE **KNOW** BILLY'S PILE IS **NORMAL-SHAPED**...

... WE CAN USE SOME **FANCY MATH** TO FIGURE OUT EXACTLY HOW THE CANS FIT INSIDE IT!

IT'S THE **CENTRAL LIMIT THEOREM!**
YAY!

I TOLD YOU THIS IS THE **MOST BEAUTIFUL SHAPE IN ALL STATISTICS!**

Average Length per Can (inches)   3   3.5   4   4.5   5

MORE SPECIFICALLY, BECAUSE THE PILE IS **NORMAL-SHAPED**...

... WE **ONLY** NEED TO KNOW ITS **CENTER VALUE** AND **STANDARD DEVIATION** TO MAKE THAT FANCY MATH WORK!*

* IF YOU LIKE MATH, SEE **PAGE 219** FOR MORE DETAILS.

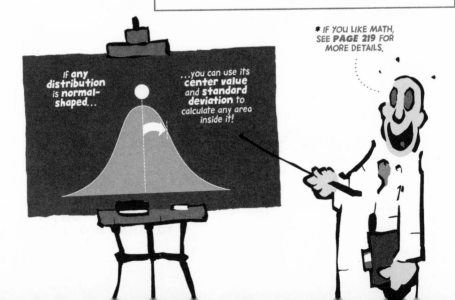

If any distribution is normal-shaped...

...you can use its **center value** and **standard deviation** to calculate any area inside it!

THE ACTUAL MATH IS **REALLY COMPLICATED.**

SO COMPLICATED THAT **STATISTICIANS DON'T EVEN DO IT!**

HELLO, COMPUTER!

FORTUNATELY, THERE ARE SOME **RULES OF THUMB** THAT WORK FOR **ANY** NORMAL DISTRIBUTION:

WE COUNT **HOW MANY STANDARD DEVIATIONS** WE ARE AWAY FROM THE CENTER.

MY PILE IS CENTERED AT **4 INCHES,** AND ITS STANDARD DEVIATION IS **0.25 INCHES.**

68% OF ALL THE CANS...

...ARE **WITHIN 1 STANDARD DEVIATION** OF THE CENTER.

IN THIS CASE THAT MEANS A RANGE OF AVERAGES **FROM 3.75 TO 4.25 INCHES.**

95% OF ALL THE CANS...

...ARE **WITHIN 2 STANDARD DEVIATIONS** OF THE CENTER.

IN THIS CASE THAT MEANS A RANGE OF AVERAGES **FROM 3.5 TO 4.5 INCHES.**

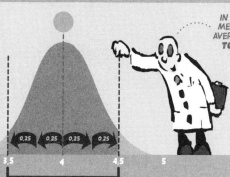

99.7% OF ALL THE CANS...

...ARE **WITHIN 3 STANDARD DEVIATIONS** OF THE CENTER.

IN THIS CASE THAT MEANS A RANGE OF AVERAGES **FROM 3.25 TO 4.75 INCHES.**

IF ALL THOSE NUMBERS FEEL **BEWILDERING...**

NUMBERS MAKE ME **NUMB,** DOC.

...JUST FOCUS ON THE **SHADED AREAS.**

CLEARLY, THERE ARE **LOTS MORE** SAMPLE CANS **INSIDE** THIS DARKER SHADED AREA OF BILLY'S PILE...

... THAN **OUT HERE,**

THE **CLUMP UNDER THE HUMP** IS MUCH BIGGER THAN THE TAILS!

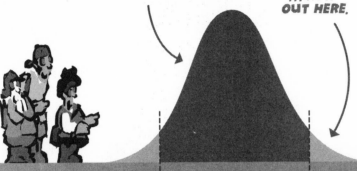

3　　3.5　　4　　4.5　　5

THE THING TO REMEMBER IS THAT THE SHADED AREAS **INSIDE BILLY'S SAMPLING DISTRIBUTION...**

... TRANSLATE DIRECTLY TO OUR CHANCES OF GRABBING AVERAGES **FROM THE SWAMP!**

THIS IS WHY STATISTICIANS **LOVE SAMPLING DISTRIBUTIONS!**

Crazy Billy's
BAIT BARN

LET'S **RECAP:**

THE **FIRST COOL THING** ABOUT BILLY'S SAMPLING DISTRIBUTION...

...IS THAT IT SHOWS US **THE POPULATION AVERAGE!**

WHAT'S THE **AVERAGE LENGTH OF ALL THE WORMS IN YOUR SWAMP,** BILLY?

THE ANSWER'S **IN MY BAIT BARN!**

THE **SECOND COOL THING** ABOUT BILLY'S SAMPLING DISTRIBUTION...

...IS THAT WE CAN USE IT TO **CALCULATE PROBABILITIES ABOUT THE OVERALL POPULATION.**

AND BECAUSE WE KNOW IT'S NORMAL...

...ALL WE NEED TO KNOW ARE ITS **CENTER VALUE...**

...AND **STANDARD DEVIATION!**

IF WE GO GRAB ANOTHER RANDOM SAMPLE OF **30 WORMS** FROM THE SWAMP...

...HOW LIKELY IS IT TO HAVE AN AVERAGE BETWEEN **3.75** AND **4.25** INCHES?

LET ME **PEER INTO MY BAIT BARN** AND TELL YOU!

CLEARLY, IF WE WERE OUT **HUNTING THE POPULATION AVERAGE**...

IT'S GOT TO BE AROUND HERE SOMEWHERE.

...A SAMPLING DISTRIBUTION LIKE THE ONE IN **CRAZY BILLY'S BAIT BARN**...

TECHNICALLY, MY **SAMPLING DISTRIBUTION**...

...IS A SPECIFIC KIND OF **PROBABILITY DISTRIBUTION**!

...WOULD BE **INCREDIBLY USEFUL** TO US.

MY PILE OF AVERAGES IS **LIKE A CRYSTAL BALL**...

...YOU CAN PEER INTO IT AND **SEE THINGS ABOUT THE POPULATION**!

ENOUGH ALREADY!

I WANT TO SEE IT!

IT'S THE MOTHER LODE!

UNFORTUNATELY...

WHA?!

IT'S EMPTY!

...IT **DOESN'T EXIST**.

IN REALITY, WE **NEVER** HAVE AN ACTUAL SAMPLING DISTRIBUTION TO LOOK AT.

IT'S ALL IN MY HEAD!

I REMEMBER EVERY CAN I'VE EVER SOLD.

IN REALITY...

...ALL WE EVER GET IS **ONE CAN**.

&#@%!

CHAPTER 9
INFERENCE

OBVIOUSLY, WE **STILL** HAVE A PROBLEM...

ANYONE GOT A **CAN OPENER?**

...AND IT BOILS DOWN TO **THIS:**

WE'RE HUNTING FOR SOMETHING THAT WE **CAN'T LOOK AT DIRECTLY.**

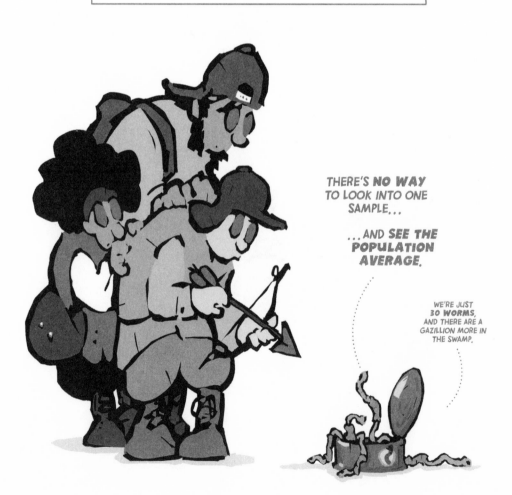

THERE'S **NO WAY** TO LOOK INTO ONE SAMPLE...

...AND **SEE THE POPULATION AVERAGE.**

WE'RE JUST **30 WORMS,** AND THERE ARE A GAZILLION MORE IN THE SWAMP.

WHEN WE MAKE A **GUESS** ABOUT THE **WHEREABOUTS OF THE POPULATION AVERAGE**...

IT'S GOT TO BE **AROUND HERE** SOMEWHERE...

...WE CAN **BASE OUR GUESS** ON **SOMETHING WE'RE CONFIDENT ABOUT**...

...I'D **PUT** MONEY ON IT.

...AND WE'VE **ALREADY LEARNED** WHAT **THAT IS:**

IN THE LONG RUN, **RANDOM SAMPLE AVERAGES** TEND TO **CLUSTER AROUND THE POPULATION AVERAGE**...

...IN **THIS BEAUTIFUL SHAPE!**

IT'S THE **CENTRAL LIMIT THEOREM!**

YAY!

HERE'S HOW TO **THINK ABOUT** WHAT WE'RE ABOUT TO DO:

BECAUSE **SAMPLE AVERAGES** TEND TO **CLUSTER**...

...**AROUND THE POPULATION AVERAGE**...

...LIKE **THIS**...

...WE CAN **DRAW A CLUSTER LIKE THIS**,...

...TO **GUESS WHERE THE POPULATION AVERAGE IS**.

I'M GUESSING IT'S **IN THE CLUMP UNDER THE HUMP!**

STATISTICIANS CALL THIS PROCESS **INFERENCE**...

WE CAN'T **SEE IT DIRECTLY**...

...SO WE LOOK FOR THE THINGS WE'D EXPECT TO **CLUSTER AROUND IT**.

IT'S LIKE **HUNTING FOR BIGFOOT**...

...BY LOOKING FOR **BIG FOOTPRINTS**.

...AND WE'RE GOING TO SPEND THE REST OF THIS CHAPTER OUTLINING **THE FIRST STEP**.

WE'RE GOING TO USE **ONE SAMPLE**...

...TO **IMAGINE WHAT WE'D SEE IF WE GATHERED A WHOLE LOT MORE**.

OUR **BASIC TASK** AT THIS STAGE...

...IS TO **DRAW A PICTURE**.

I TOLD YOU STATISTICS WAS ALL ABOUT **DRAWING PICTURES!**

IT'S A PICTURE OF HOW WE THINK **SAMPLE AVERAGES WOULD PILE UP**...

...**IF** WE WENT OUT AND **GATHERED GAZILLIONS OF THEM**.

REMEMBER, WE DRAW THIS THING...

...BECAUSE WE'RE **HUNTING FOR THE POPULATION AVERAGE**.

YOO HOO, **ARE YOU HIDING IN THERE?**

AND IT'S ALL BASED ON INFORMATION WE GET FROM OUR **ONE RANDOM SAMPLE**.

WHAT'S YOUR **SAMPLE SIZE**...

...AND YOUR **LOCATION**...

...AND YOUR **SPREAD?**

IT'S LIKE WE'RE MAKING OUR **BEST POSSIBLE GUESS** AT WHAT CRAZY BILLY'S IMAGINARY PILE LOOKS LIKE...

... USING **ONLY ONE CAN**.

THIS MIGHT **SEEM LIKE MAGIC**...

I PUT **ONE CAN** OF WORMS INTO THE HAT.

NOTHING UP MY SLEEVES.

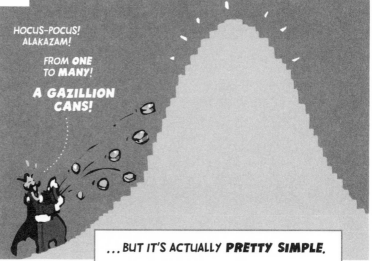

HOCUS-POCUS! ALAKAZAM!

FROM **ONE** TO **MANY**!

A GAZILLION CANS!

... BUT IT'S ACTUALLY **PRETTY SIMPLE**.

TO DRAW OUR PICTURE, WE USE THE CENTRAL LIMIT THEOREM AS A **BLUEPRINT:**

WE CAN EXPECT GIANT PILES OF SAMPLE AVERAGES TO BE **NORMAL**...[1]

...AND **CENTERED AT THE POPULATION AVERAGE**...

...WITH A **STANDARD DEVIATION** EQUAL TO...

...**THE POPULATION STANDARD DEVIATION** DIVIDED BY THE SQUARE ROOT OF THE SAMPLE SIZE.

WHEW!

1. REMEMBER, SOME **RESTRICTIONS APPLY**. SEE PAGE 102.

HOWEVER, BECAUSE WE **DON'T ACTUALLY KNOW THE REAL POPULATION VALUES**...

YOU **NEVER** WILL!

...WE SIMPLY **REPLACE THEM** WITH VALUES WE GRAB **FROM OUR SAMPLE**.

PRETEND FOR A MOMENT THAT THE AVERAGE IN YOUR CAN **IS THE SAME AS THE AVERAGE IN THE SWAMP!**

AND THE SPREAD IN YOUR CAN **IS THE SAME AS THE SPREAD IN THE SWAMP!**

ISN'T THAT **CHEATING!?**

NOPE, JUST OUR **BEST GUESS**.

WE LIKE TO CALL IT **APPROXIMATION**.

SO, FOR EXAMPLE, WHEN WE USE
SAMPLE VALUES FROM OUR ONE CAN...

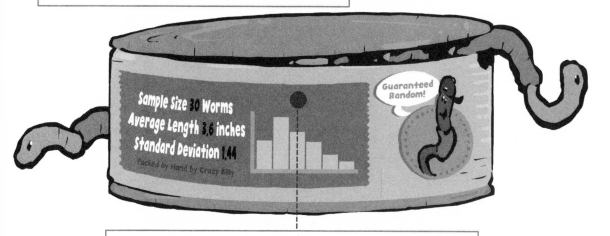

Sample Size **30** Worms
Average Length **3.6** inches
Standard Deviation **1.44**
Packed by Hand by Crazy Billy

Guaranteed Random!

... **WE DRAW A PICTURE THAT LOOKS LIKE THIS:**

3.6

OUR **ESTIMATED** GIANT PILE OF AVERAGES IS **NORMAL**...

...AND **CENTERED AT OUR ONE CAN'S AVERAGE**...

... WITH A **STANDARD DEVIATION** EQUAL TO ...

...**OUR CAN'S STANDARD DEVIATION** DIVIDED BY THE SQUARE ROOT OF THE SAMPLE SIZE!

WHEW!

$$SD = \frac{1.44}{\sqrt{30}}$$

WE CALL THIS PICTURE AN **ESTIMATED SAMPLING DISTRIBUTION.***

IT'S AN **ESTIMATE**...

...OF HOW **SAMPLE AVERAGES** WOULD BE **DISTRIBUTED**...

...IF WE COLLECTED TONS OF THEM.

* SEE **PAGE 219** TO LEARN TO DESCRIBE THIS USING MATH SYMBOLS.

CHAPTER 10
CONFIDENCE

REMEMBER THAT OUR ULTIMATE GOAL IS TO **LEARN SOMETHING ABOUT THE POPULATION AVERAGE.**

I DON'T ACTUALLY CARE ABOUT **BIG** FOOTPRINTS...

...I WANT TO KNOW ABOUT **BIGFOOT HIMSELF!**

UNFORTUNATELY, DESPITE THE **MAGIC TRICKS** WE JUST LEARNED...

NOTHING UP MY SLEEVES...

...EXCEPT A CAN OF WORMS.

... WE WILL STILL **NEVER** BE ABLE TO **PINPOINT IT DIRECTLY...**

THERE'S **NO WAY** TO PEER INTO ONE CAN AND **SEE THE POPULATION AVERAGE...**

...AND THERE'S **NO WAY** TO PEER INTO ONE ESTIMATED **SAMPLING DISTRIBUTION** AND SEE IT EITHER.

...EVER!

SIGH.

THAT'S WHY WE'RE LEARNING TO **DO INFERENCE.**

I'LL **NEVER** BE ABLE TO FIND WHAT I'M LOOKING FOR!

DON'T DESPAIR!

YOU CAN STILL **MAKE A GOOD GUESS** ABOUT ITS WHEREABOUTS!

SO FAR WE'VE COVERED THE **FIRST STEP** IN THE **INFERENCE PROCESS...**

...BUT WE STILL NEED TO COVER **THE SECOND.**

CAREFULLY, LOVINGLY DRAW YOUR ESTIMATED SAMPLING DISTRIBUTION...

...WITH EXQUISITE ATTENTION TO THE IMPORTANT AND CRUCIAL DETAILS.

NOW CHOP IT INTO BITS!

SO IN THIS CHAPTER WE'RE GOING TO LEARN TO **REFINE OUR DRAWING...**

...GIVE IT A LITTLE **TRIM** AROUND THE EDGES...

...AND USE WHAT'S LEFT TO **CALCULATE OUR CONFIDENCE.**

MUCH BETTER!

NOW I CAN BE **CONFIDENT** ABOUT WHAT I'M LOOKING AT.

SINCE WE ALREADY KNOW **HOW TO DRAW AN ESTIMATED SAMPLING DISTRIBUTION...**

ISN'T SHE BEAUTIFUL?

...LEARNING HOW TO CALCULATE OUR CONFIDENCE IS EASY.

WE SIMPLY **PEER DOWN INTO** THE THING WE JUST DREW...

THIS TIME LET'S MEASURE **2 STANDARD DEVIATIONS AWAY FROM THE CENTER VALUE...**

...ON EITHER SIDE.

...AND CHOP OFF ITS TAILS!

SNIP SNIP!

HOLD STILL, THIS WON'T HURT A BIT.

THEN **WE MAKE A STATEMENT LIKE THIS ONE:**

WE'RE **95%** CONFIDENT...

...THAT THE **POPULATION AVERAGE IS SOMEWHERE INSIDE THIS RANGE!**

YAY!

WE'LL RETURN TO THESE BITS IN **CHAPTER 11.**

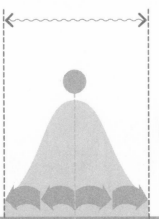

THAT'S **ALL IT TAKES**.

MEASURE AND *CHOP!*

ANYONE CAN DO IT!

TECHNICALLY, IF WE WANT TO BE **MORE CONFIDENT**, WE CAN **CHOP FARTHER OUT**.

IF WE WANT TO BE 99.7% CONFIDENT...

...WE MEASURE AND CHOP 3 SDs AWAY FROM THE CENTER...

...ON EITHER SIDE.

AND IF WE WANT TO BE **LESS CONFIDENT**, WE CAN **CHOP CLOSER IN**.

IF WE WANT TO BE 68% CONFIDENT...

...WE MEASURE AND CHOP 1 SD AWAY FROM THE CENTER...

...ON EITHER SIDE.

IT'S BASED ON THE RULES WE LEARNED ON **PAGE 115**!

BUT WHEREVER WE CHOP, WE **ALWAYS** DECLARE OUR CONFIDENCE WITH A **TWO-PART STATEMENT**, LIKE THIS...

WE'RE **95%** CONFIDENT...

...THAT THE POPULATION AVERAGE IS SOMEWHERE INSIDE **THIS** RANGE!

...THAT COMBINES A PARTICULAR **LEVEL OF CONFIDENCE**...

...WITH A **CONFIDENCE INTERVAL**.*

THAT'S **THE LONG-RUN ODDS THAT WE'RE RIGHT**!

WE'LL **NEVER** ACTUALLY CATCH IT...

...BUT OUR BEST BET IS THAT IT'S **IN HERE**!

* TECHNICALLY, A CONFIDENCE INTERVAL IS A KIND OF **INTERVAL ESTIMATE**. SEE **PAGE 220** FOR MORE.

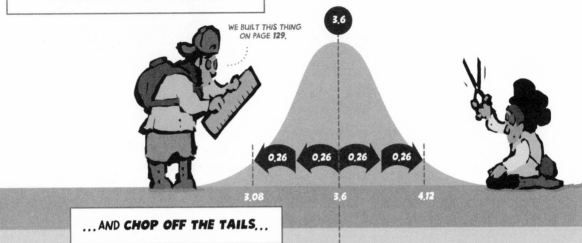

FOR EXAMPLE, IF WE TAKE THE ESTIMATED SAMPLING DISTRIBUTION WE BUILT WITH OUR **ONE CAN OF WORMS...**

WE BUILT THIS THING ON PAGE **129.**

3.6

0.26 0.26 0.26 0.26

3.08 3.6 4.12

...AND **CHOP OFF THE TAILS...**

...AT A DISTANCE **2 STANDARD DEVIATIONS** AWAY FROM THE CENTER VALUE...

3.08 4.12

... WE CAN SAY **THIS:**

WE'RE **95%** CONFIDENT...

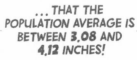

... THAT THE POPULATION AVERAGE IS BETWEEN **3.08** AND **4.12** INCHES!

BUT WHAT **EXACTLY DOES IT MEAN?**

WE JUST BUILT **ONE** ESTIMATED SAMPLING DISTRIBUTION WITH **ONE** RANDOM SAMPLE...

WE BUILT THIS WITH **ONE CAN** OF 30 WORMS.

3.6

0.26

...AND USED IT TO CALCULATE **ONE** CONFIDENCE INTERVAL.

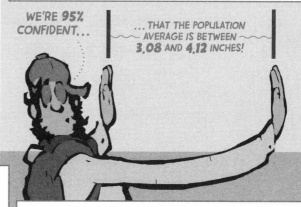

WE'RE **95%** CONFIDENT...

...THAT THE POPULATION AVERAGE IS BETWEEN **3.08** AND **4.12** INCHES!

BUT IF WE GRABBED **ANOTHER RANDOM SAMPLE** AND USED IT TO BUILD **ANOTHER ESTIMATED SAMPLING DISTRIBUTION**...

WE BUILT THIS WITH A **DIFFERENT CAN** OF 30 WORMS.

IT HAS A DIFFERENT CENTER VALUE...

...AND A DIFFERENT SPREAD.

4.1

0.23

...WE'D MOST LIKELY GET A **DIFFERENT INTERVAL!**

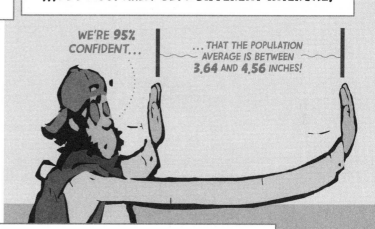

WE'RE **95%** CONFIDENT...

...THAT THE POPULATION AVERAGE IS BETWEEN **3.64** AND **4.56** INCHES!

AND **IF** WE KEPT ON GATHERING **MORE** RANDOM SAMPLES AND BUILDING **MORE** ESTIMATED SAMPLING DISTRIBUTIONS...

GATHER AND ESTIMATE.

GATHER AND ESTIMATE.

GATHER AND ESTIMATE.

GATHER AND ESTIMATE.

GATHER AND ESTIMATE.

GATHER AND ESTIMATE.

GATHER AND ESTIMATE.

GATHER AND ESTIMATE.

GATHER AND ESTIMATE.

...WE'D KEEP ON GETTING **DIFFERENT INTERVALS.**

137

THIS IS IMPORTANT BECAUSE THE **ONLY THING** THAT THIS ACTUALLY MEANS...

WE'RE **95%** CONFIDENT...

... THAT THE POPULATION AVERAGE IS SOMEWHERE INSIDE **THIS** RANGE!

... IS THAT **IF** WE CALCULATED A GAZILLION DIFFERENT RANGES **IN THIS SAME WAY**...

GRAB RANDOMLY...

... DRAW PICTURE...

... CHOP OFF TAILS **2 SDs** AWAY FROM THE CENTER.

... ABOUT 19 OUT OF EVERY 20 **WOULD** INCLUDE THE TRUE POPULATION AVERAGE...

19 OUT OF **20** IS 95%!

... AND ABOUT 1 OUT OF EVERY 20 **WOULDN'T**.

1 OUT OF 20 IS **COMPLETELY WRONG!**

LET'S HOPE WE DIDN'T GRAB **THAT** CAN.

IN OTHER WORDS, WHEN WE SAY **THIS**:

WE'RE **95%** CONFIDENT...

...THAT THE POPULATION AVERAGE IS SOMEWHERE INSIDE **THIS** RANGE!

...IT MEANS THERE'S A **5% CHANCE** THAT WE'RE **WRONG ABOUT THIS.**

IN WHICH CASE, THE POPULATION AVERAGE IS **REALLY SOMEWHERE ELSE**...

...AND WE **MISSED THE MARK ENTIRELY!**

THE **SAD TRUTH** IS THAT **ANY ONE SAMPLE** WE ASSEMBLE RANDOMLY FROM A POPULATION...

...28 WORMS...

...29 WORMS...

...30 WORMS!

...MIGHT BE MISLEADING.

WE **MIGHT** GRAB 30 INCREDIBLY SHORT WORMS...

...BY CHANCE.

AND **IF OUR ONE SAMPLE** IS MISLEADING...

...THE ESTIMATED SAMPLING DISTRIBUTION WE BUILD ON TOP OF IT **WILL BE MISLEADING, TOO.**

IF WE BUILD THIS DISTRIBUTION WITH 30 **INCREDIBLY SHORT WORMS...**

...IT WILL BE **TOO FAR TO THE LEFT.**

WHAT IF THE POPULATION AVERAGE **IS REALLY OVER HERE SOMEWHERE?**

IT'S A **SERIOUS CONCERN...**

...BUT WE CAN GET AROUND IT BY **ALWAYS** KEEPING THE **BIGGER PICTURE** IN MIND.

THIS CAN OF WORMS **COULD BE A DUD!**

MAYBE, BUT **PROBABLY NOT!**

THINK ABOUT THE **LONG RUN!**

IN THE LONG RUN, DESPITE THE FACT THAT OUR ONE SAMPLE **MIGHT** BE MISLEADING...

WE **MIGHT** GRAB 30 INCREDIBLY WEIRD WORMS...

...BY CHANCE!

...IT **PROBABLY ISN'T**...

...BECAUSE **MOST RANDOM SAMPLE AVERAGES** TEND TO CLUSTER AROUND THE POPULATION AVERAGE!

SOUND FAMILIAR?

IT'S THE **CENTRAL LIMIT THEOREM!**

YAY!

IN OTHER WORDS, OUR ONE CAN **MIGHT** RANDOMLY HAVE AN AVERAGE VALUE **WAY OUT HERE**...

WEIRDLY SHORT WORMS.

...OR **WAY OUT HERE**...

WEIRDLY LONG WORMS.

...BUT IT'S **UNLIKELY**...

PRETTY NORMAL WORMS.

IN THE LONG RUN, MOST CANS HAVE AVERAGES **IN THE CLUMP UNDER THE HUMP.**

...AND WE CAN **BE CONFIDENT** ABOUT THAT.

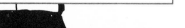

IN SUM, UNDERSTANDING **STATISTICAL CONFIDENCE**...

WE'RE **95%** CONFIDENCE...

...THAT THE POPULATION AVERAGE IS SOMEWHERE INSIDE **THIS** RANGE!

...INVOLVES KEEPING OUR EYES ON THE LONG RUN **AND** THE SHORT RUN **AT THE SAME TIME.**

IN THE LONG RUN, OUR ESTIMATION AND CHOPPING CALCULATIONS WORK, **PERIOD**...

IF YOU GRAB A GOOD-SIZED **RANDOM SAMPLE** AND USE IT TO **BUILD AN ESTIMATED SAMPLING DISTRIBUTION**...

...THEN MEASURE 2 SDs **FROM THE CENTER** AND **CHOP OFF THE END BITS**...

...**95% OF THE TIME** YOU'LL GET A RANGE THAT INCLUDES THE **TRUE POPULATION AVERAGE!**

IT'S BEEN **PROVEN WITH MATH!**

AND **EXPERIENCE!**

...BUT **IN THE SHORT RUN,** IT'S ALWAYS POSSIBLE THAT WE **GRABBED A DUD!**

WE'RE 95% **CONFIDENT** THAT THE POPULATION AVERAGE IS **SOMEWHERE INSIDE THIS RANGE**...

...**BUT IS IT REALLY?**

IT IS **OR** IT ISN'T.

WE HAVE **NO WAY OF KNOWING FOR CERTAIN!**

CHAPTER 11
THEY HATE US

GRRRRR.

THEY **WANT TO KILL** US!

HOW **CONFIDENT** ARE YOU ABOUT THAT?

WHEN WE USE ONE **SAMPLE**...

BASED ON THESE **50** RANDOM MERMAIDS...

...TO **CALCULATE OUR STATISTICAL CONFIDENCE** ABOUT AN ENTIRE **POPULATION**...

...I'M 95% CONFIDENT...

...THAT ALL THE MERMAIDS IN THIS ENTIRE LAGOON...

...AVERAGE BETWEEN **7** AND **10** CENTIMETERS TALL!

...WE'RE DOING SOMETHING **AMAZING**!

WE'RE MAKING A **GUESS**...

...A **DEPENDABLE** GUESS...

WHO WOULD HAVE THOUGHT MERMAIDS WERE **SO SMALL**?

ONE RANDOM SAMPLE MIGHT BE MISLEADING...

...BUT IN THE LONG RUN THAT'S **DOUBTFUL**.

...ABOUT **SOMETHING WE CAN'T SEE,**
BUT CAN ONLY IMAGINE.

YOO HOO...

...ARE THERE
MORE OF YOU
DOWN THERE?

HE'LL **NEVER**
KNOW FOR
CERTAIN.

IT ALL STARTS WITH ONLY **THREE NUMBERS**.

A REASONABLY LARGE **SAMPLE SIZE**...

...A SAMPLE **AVERAGE**...

...AND A SAMPLE **STANDARD DEVIATION**.

BUT REMEMBER, IT ONLY WORKS IF YOU GATHER YOUR SAMPLE MERMAIDS **RANDOMLY**.

WITH ONLY THESE THREE NUMBERS WE CAN **BUILD AN ESTIMATED SAMPLING DISTRIBUTION**...

WHICH IS LIKE **IMAGINING** HOW A GAZILLION SAMPLES WOULD LOOK...

...IF WE PILED THEM UP BY THEIR AVERAGE VALUE.

8.5

1.35

WE KNOW THE PILE WOULD BE **NORMAL-SHAPED** IF THE SAMPLE SIZE IS LARGE ENOUGH.

PLUS, WE CAN ESTIMATE ITS CENTER VALUE AND STANDARD DEVIATION WITH OUR SAMPLE VALUES.

5.5 6 6.5 7 7.5 8 8.5 9 9.5 10 10.5 11 11.5

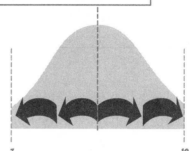

...AND **CHOP OFF ITS TAILS**...

WE COUNT OUTWARD BY ITS **STANDARD DEVIATIONS**...

...TO LEAVE THIS **MIDDLE CLUMP** WHOSE **PROBABILITY VALUE** WE KNOW PRECISELY.

7   10

...TO GET A **SINGLE DEPENDABLE STATEMENT**...

IN THE **LONG RUN**, THIS WORKS!

...THAT COMBINES A **LEVEL OF CONFIDENCE**...

...WITH A **CONFIDENCE INTERVAL**.

I'M **95%** CONFIDENT...

...THAT THE POPULATION AVERAGE OF ALL THE MERMAIDS IN THE ENTIRE LAGOON IS **SOMEWHERE** BETWEEN 7 AND 10 CENTIMETERS!

AS WE'VE LEARNED, BECAUSE THIS PROCESS INVOLVES A BUNCH OF **MATH**...

AAHHHHH!

...IT WORKS ONLY ON **QUALITIES** THAT WE **CAN MEASURE WITH NUMBERS.**

HOW MUCH DO MERMAIDS **WEIGH?**

HOW **LONG** ARE THEY?

HOW MANY **TEETH** DO THEY HAVE?

YOU NEED **NUMERICAL DATA.**

WE COVERED THAT IN **CHAPTER 4.**

SO IT **MIGHT SEEM** LIKE IT **WON'T WORK** ON QUALITIES THAT **AREN'T OBVIOUSLY NUMERICAL**...

ARE THEY **HAPPY?**

DO THEY ALL **SING BEAUTIFULLY?**

IF I POKE THEM WITH A STICK, **HOW MUCH DO THEY HURT?**

...BUT THAT'S **NOT ALWAYS THE CASE.**

THE TRUTH IS, WE CAN CALCULATE OUR CONFIDENCE ABOUT **ANY QUALITY**...

ARE MERMAIDS **OPTIMISTIC?**

HOW **SMART** ARE THEY?

DO THEY ENJOY EATING **SUSHI?**

... **IF** WE CAN FIGURE OUT A WAY TO **MEASURE THAT QUALITY**...

TAKE THIS **TEST.**

... AND ARRANGE IT ON A **NUMBER LINE.**

IF YOU SCORED OVER HERE, YOU'RE AN **IDIOT!**

IF YOU SCORED OVER HERE, YOU'RE A **GENIUS!**

Score: 60   80   100   120   140

IN THIS CHAPTER WE'RE GOING TO DO JUST THIS SORT OF THING...

... TO INVESTIGATE A QUESTION ABOUT **HATRED.**

HOW DO I **HATE THEE?**

LET ME **COUNT THE WAYS...**

149

EVERYONE KNOWS THAT THE AWFUL WRETCHES ON THE PLANET OF BLIP...
...HATE THE GOOD PEOPLE ON THE NEIGHBORING PLANET OF BLEEEP.
NYA NYA!
YOU @*$&-ING BLEEEPS!
THEY HATE US!
THEY'VE HATED US FOR A TRILLION GENERATIONS.
AND OUR QUESTION IS...
...IS THIS TRUE?
BECAUSE WE CAN'T POSSIBLY ASK ALL THE 785,000,000,000 BLIPS WHO LIVE ON THAT PLANET HOW THEY FEEL...
...WE HAVE NO CHOICE BUT TO BASE OUR ANSWER ON A RANDOM SAMPLE OF THEM.
THAT'S A LOT OF BLIPS.
IN OTHER WORDS, WE HAVE TO USE STATISTICS.
REMEMBER, NOBODY USES STATISTICS UNLESS THEY HAVE TO!

HOWEVER, BEFORE WE CAN GO OUT AND **GATHER A RANDOM SAMPLE**...

...WE HAVE TO FIGURE OUT A WAY TO TRANSLATE **HOW EACH BLIP FEELS ABOUT BLEEEPS**...

...INTO A **NUMBER**.

I LIKE TO **DECAPITATE** BLEEEPS!

WE NEED A NUMBER FOR THAT.

DON'T TELL ANYONE...

...BUT **I THINK THEY'RE CUTE!**

LET'S MAKE THAT A **MORE POSITIVE NUMBER!**

IN THIS CASE WE CAN **INVENT A SCALE**...

...THAT RANGES FROM **PURE HATRED**...

...TO PURE **LOVE**.

I WANT TO **KILL EVERY BLEEEP I SEE**.

I DON'T THINK I LIKE THEM VERY MUCH.

TO BE HONEST, I'M **INDIFFERENT** ABOUT THEM.

THEY'RE **SORT OF COOL**, I GUESS.

MY DREAM IS TO **MARRY A BLEEEP!**

-10   -5   0   5   10

SO, FOR EACH BLIP WE INTERVIEW...

...WE'LL GET **A SINGLE NUMBER BETWEEN -10 AND 10.**

I'D LIKE TO **CRUNCH THEIR BLEEEPING BONES** AND **BURP SARCASTICALLY!**

OKAY, YOU'RE A **-10!**

THE NEXT STEP IS TO PAINSTAKINGLY **GATHER OUR SAMPLE DATA**...

... **RANDOMLY**...

THERE CAN BE **NO SYSTEMATIC DIFFERENCE**...

... BETWEEN THE SAMPLE OF BLIPS WE INTERVIEW...

... AND ANY OTHER SAMPLE WE MIGHT GRAB...

... SO LET'S GRAB OUR SAMPLE BLIPS RANDOMLY **FROM ALL OVER THEIR PLANET**...

... FROM **HERE**...

... BACK **HERE**...

...**HERE**...

...**HERE**...

...**HERE**...

...**HERE**...

...**HERE**...

...**ETC**... ...**ETC**...

... KEEPING IN MIND THAT WE HAVE TO **GATHER ENOUGH OF IT**...

... TO **MAKE SURE THE MATH WORKS**.

... WE WON'T STOP UNTIL WE'VE INTERVIEWED **100 OF THEM**.

THEN, ONCE WE'VE ASSEMBLED
OUR **SAMPLE DATA**...

OVERALL, OUR DATA
IS SLIGHTLY **SHIFTED**
LEFT OF ZERO...

...AND **SKEWED** TO
THE RIGHT.

THESE FIVE BLIPS
TRULY **HATE**
BLEEEPS.

ONE BLIP
REPORTED
LOVING
BLEEEPS!

Blip Frequency

Sentiment
Toward
Bleeeps

... WE CAN **PONDER IT A BIT**...

CLEARLY, LOTS OF OUR
SAMPLE BLIPS FEEL
NEGATIVELY ABOUT
BLEEEPS...

...BUT DOES THAT MEAN
THE **WHOLE PLANET**
IS **SEETHING WITH
HATRED?**

... AS WE EXTRACT THE **THREE NUMBERS** THAT WE NEED TO DO **STATISTICAL INFERENCE.**

THE **SAMPLE
SIZE IS 100.**

LARGE ENOUGH
TO CALCULATE
OUR CONFIDENCE.

THE **SAMPLE
AVERAGE IS -1.**

NEGATIVE, BUT NOT
SEVERELY SO.

THE **SAMPLE
SD IS 4.**

SEEMS LIKE **A LOT OF
VARIATION**, CONSIDERING
THAT THE ENTIRE SCALE IS
ONLY 20 UNITS WIDE.

HMMMM.

NOW WHAT CAN
WE SAY ABOUT ALL
785,000,000,000 OR
SO BLIPS IN THE ENTIRE
POPULATION?

153

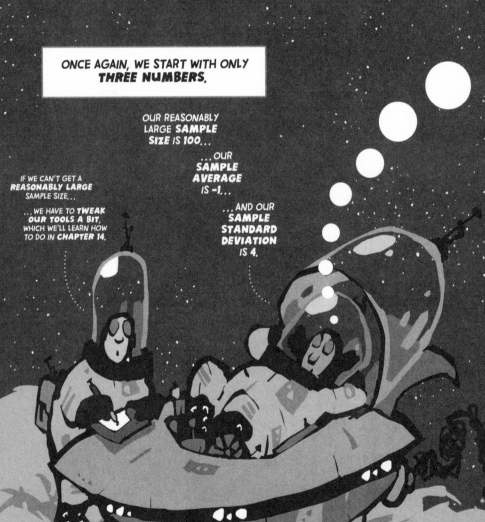

ONCE AGAIN, WE START WITH ONLY **THREE NUMBERS.**

OUR REASONABLY LARGE **SAMPLE SIZE** IS 100...

...OUR **SAMPLE AVERAGE** IS −1...

IF WE CAN'T GET A **REASONABLY LARGE** SAMPLE SIZE...

...WE HAVE TO **TWEAK OUR TOOLS A BIT,** WHICH WE'LL LEARN HOW TO DO IN **CHAPTER 14.**

...AND OUR **SAMPLE STANDARD DEVIATION** IS 4.

SO DO BLIPS...

...HATE BLEEEPS?

WE WANT TO **MANGLE** YOUR EYEBALLS WITH OUR THUMBS...

...CUT OFF YOUR TAILS WITH A DULL KNIFE...

...REGURGITATE YOUR SPLEEN!

...THEN LIGHT THE WHOLE THING **ON FIRE!**

SEE...? THEY'RE A BUNCH OF **SPITEFUL EVILDOERS!**

SINCE WE CAN'T INTERVIEW THE ENTIRE POPULATION, WE'LL **NEVER KNOW FOR CERTAIN.**

BUT OUR RANDOM SAMPLE DATA...

...**STRONGLY** SUGGESTS THEY **DON'T,** ACTUALLY.

WE'RE **95%** CONFIDENT...

...THAT THE TRUE POPULATION AVERAGE ISN'T OVER THERE IN THE **HATE ZONE**...

...BUT SOMEWHERE IN **HERE!**

-1.8   -0.2

THAT'S **MILD** DISLIKE AT BEST...

...VERGING ON **INDIFFERENCE**...

...BUT **NOT HATE!**

-10   -5   0   5   10

REMEMBER THAT THE **BEST WE CAN OFFER** WITH STATISTICS IS A **NUANCED PORTRAIT**...

KEEP YOUR EYES ON THE **LONG TERM** AND THE **SHORT TERM**...

... AT THE SAME TIME!

... BECAUSE **ANY CONCLUSION BASED ON ONE SAMPLE**...

WHICH MEANS **ALL** CONCLUSIONS MADE WITH STATISTICS,

ANYWHERE, ANYTIME, EVER!

... **MIGHT BE DEAD WRONG.**

WHEN WE'RE **95%** CONFIDENT IT'S IN HERE...

... WE'RE ALSO **5%** CONFIDENT THAT **IT'S NOT!**

AND EVEN IF WE **EXPAND OUR LEVEL OF CONFIDENCE**...

... TO COVER A **WIDER** INTERVAL...

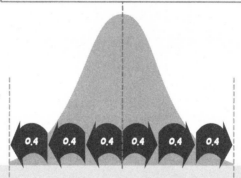

IF WE COUNT **3** STANDARD DEVIATIONS OUTWARD, WE CAN SAY:

0.4 0.4 0.4 0.4 0.4 0.4

-2.2   -1   0.2

WE'RE **99.7%** CONFIDENT...

... THAT THE POPULATION AVERAGE OF ALL THE BLIPS ON THE ENTIRE PLANET IS ~ SOMEWHERE BETWEEN ~ -2.2 AND 0.2!

... WE STILL **MIGHT** BE WRONG.

OUR SAMPLE COULD BE **MISLEADING**...

... IT'S JUST **VERY UNLIKELY.**

BUT THERE'S **ONE MORE THING** TO MENTION...

WE JUST CALCULATED **95%** AND **99.7%** CONFIDENCE INTERVALS...

...BASED ON ONE RANDOM SAMPLE OF **100 BLIPS.**

WE'RE **REALLY CONFIDENT**...

...THAT WE PROBABLY **DON'T LIKE YOU VERY MUCH!**

BUT THERE'S ONE OTHER THING WE COULD HAVE DONE TO **BUY OURSELVES MORE CONFIDENCE.**

YOU CAN **NEVER GET ENOUGH CONFIDENCE.**

MORE IS **ALWAYS BETTER!**

IF WE HAD STARTED THE WHOLE PROCESS BY INTERVIEWING **MORE RANDOM BLIPS**...

ON A SCALE OF -10 TO 10, HOW DO **YOU** FEEL ABOUT BLEEEPS?

LET'S INTERVIEW **MORE THAN 100!**

LET'S NOT STOP UNTIL WE'VE INTERVIEWED **225!**

...OUR ESTIMATED SAMPLING DISTRIBUTION WOULD HAVE BEEN **NARROWER**...

LOOK WHAT HAPPENS WHEN WE **INCREASE THE SAMPLE SIZE** FROM **100** TO **225.**

THE WHOLE PILE BECOMES **MORE SLENDER!**

WE PREDICTED THIS WOULD HAPPEN ON **PAGE 98.**

-1

$\frac{4}{\sqrt{100}}$

0.4

-1

$\frac{4}{\sqrt{225}}$

0.26

...AND THAT WOULD HAVE MADE OUR CONFIDENCE INTERVALS **NARROWER** AND THEREFORE **MORE PRECISE!**

TO SEE HOW IT WORKS, IMAGINE WE HAD ORIGINALLY STARTED WITH **THESE THREE NUMBERS.**

IT'S **UNLIKELY** OUR SAMPLE SIZE AND SAMPLE SD WOULD **REALLY BE THE SAME** WITH A **LARGER SAMPLE**...

...BUT LET'S USE THE SAME NUMBERS SO WE CAN SEE THE EFFECTS OF CHANGING **ONLY THE SAMPLE SIZE.**

THE **SAMPLE SIZE IS 225.**

THAT'S MORE THAN TWICE AS MANY BLIPS.

THE **SAMPLE AVERAGE IS -1.**

THE **SAMPLE SD IS 4.**

JUST LIKE BEFORE, WITH THESE THREE NUMBERS WE CAN **BUILD AN ESTIMATED SAMPLING DISTRIBUTION...**

IT'S NORMAL...

...AND CENTERED AT OUR SAMPLE AVERAGE...

...BUT THE WHOLE THING IS **NARROWER** THIS TIME BECAUSE OUR SAMPLE SIZE WAS LARGER.

-1

$\dfrac{4}{\sqrt{225}}$

...AND **CHOP OFF ITS TAILS.**

LET'S COUNT 3 STANDARD DEVIATIONS OUTWARD TO GET A **99.7% CONFIDENCE INTERVAL.**

| 0.26 | 0.26 | 0.26 | 0.26 | 0.26 | 0.26 |

-2.5   -2.0   -1.5   -1.0   -0.5   0.0   0.5

AND THIS TIME **ANY PARTICULAR CONFIDENCE LEVEL** WE CHOOSE...

... WILL HAVE A **MUCH MORE PRECISE** INTERVAL.

THIS TIME WE'RE **99.7% CONFIDENT...**

... THAT THE POPULATION AVERAGE IS **SOMEWHERE BETWEEN** -1.78 AND -0.22.

THAT'S ABOUT THE SAME AS THE 95% CONFIDENCE INTERVAL WE GOT WITH 100 BLIPS!

THIS, ULTIMATELY, IS THE REASON A **LARGER SAMPLE SIZE** IS BETTER!

IF YOU CAN **INCREASE YOUR SAMPLE SIZE...**

...YOU **SHOULD.**

IT'LL MAKE YOU **MORE CONFIDENT!**

IN THIS CHAPTER WE CREATED A **NUMERICAL SCALE** THAT DESCRIBES **HOW ONE GROUP OF PEOPLE FEELS ABOUT ANOTHER.**

I'M A **10!**

... WILL YOU **MARRY ME?**

THIS SAME SORT OF TRICK CAN WORK TO INVESTIGATE **ALL KINDS OF DIFFERENT QUESTIONS...**

ON A SCALE OF **1 TO 10**... ...TELL ME HOW MUCH THIS **HURTS.**

ON A SCALE OF **0 TO -100**... ...HOW **GRUMPY** ARE YOU IN THE MORNINGS?

ON A SCALE OF **1 TO 100,000,000,000,000**... ...HOW MUCH DO YOU **LOVE ME?**

...BECAUSE **IF** WE CAN GATHER ENOUGH **RANDOM NUMERICAL SAMPLE DATA...**

...WE CAN **CALCULATE OUR CONFIDENCE** ABOUT **ANY POPULATION...**

DO YOU FIND ME **ATTRACTIVE?**

... THAT **LIES BEYOND OUR REACH.**

CHAPTER 12
HYPOTHESIS TESTING

IN THE LAST FEW CHAPTERS WE LEARNED HOW TO **CALCULATE OUR CONFIDENCE...**

MEASURE AND **CHOP!**

IN THE LONG RUN, IT **WORKS.**

...BY BUILDING AN ESTIMATED SAMPLING DISTRIBUTION...

WE MADE THIS WITH **ONLY ONE SAMPLE!**

IT'S A GUESS ABOUT HOW A **GAZILLION SAMPLES** WOULD PILE UP IF WE ACTUALLY COLLECTED THEM ALL...

...AND SORTED THEM BY THEIR AVERAGES!

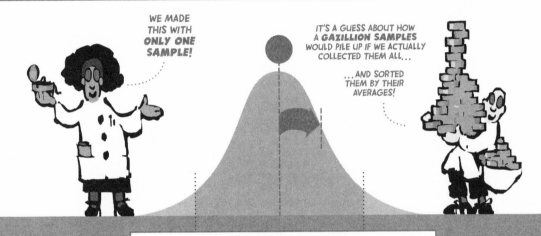

...AND CARVING OUT A BIG CLUMP IN THE MIDDLE OF IT.

WE'RE **95% CONFIDENT...**

...THAT THE POPULATION AVERAGE IS **SOMEWHERE INSIDE THAT CLUMP!**

IN THIS CHAPTER WE'RE GOING TO LEARN A **NEW TECHNIQUE**.

WE'RE GOING TO **TAKE OUR ESTIMATE**...

...AND SEE WHAT IT CAN TELL US WHEN WE **SLIDE IT TO A NEW CENTER LOCATION**.

IT'S ALL PART OF A PROCESS CALLED **HYPOTHESIS TESTING**...*

THIS IS A **TEST**...

...OF WHETHER WE THINK THE POPULATION AVERAGE **MIGHT ACTUALLY REALLY BE**...

...**RIGHT HERE!!**

...ANOTHER **MAJOR STRATEGY** FOR **DOING INFERENCE**.

* SEE **PAGE 221** FOR A TECHNICAL DESCRIPTION OF THE STUFF WE COVER IN THIS CHAPTER.

HYPOTHESIS TESTING **SOUNDS FANCY...**

THEY CALL ME **REGINALD HYPOTHESIS JONES THE 3RD...**

...AND **MY SOCKS COST MORE THAN YOUR JACKET.**

...BUT IT'S REALLY JUST ANOTHER WAY TO **HUNT FOR THE ELUSIVE POPULATION AVERAGE.**

SHSHSH.

AS WE'VE LEARNED, WE'LL **NEVER** BE ABLE TO SEE THE REAL POPULATION AVERAGE DIRECTLY...

BLINDFOLDS **ON!**

...BUT IT TURNS OUT WE CAN MAKE SOME HEADWAY IN OUR QUEST BY **TAKING A STAB AT ITS PRECISE LOCATION.**

HEY, FELLAS, WHAT IF, JUST MAYBE, IT WAS **EXACTLY RIGHT HERE!**

RIGHT UNDER OUR NOSES!

WHAT WE'LL ULTIMATELY BE DOING WITH **HYPOTHESIS TESTING**...

...IS **TESTING OUR GUESS**...

...BY **COMPARING IT** TO THE AVERAGE IN THE SAMPLE **WE ACTUALLY FOUND**.

I THINK **THIS** COULD BE IT!

WHAT IF THIS IS IT?

IF THAT'S **REALLY** IT, WHAT DO YOU MAKE OF THIS, EH?

BUT THE WHOLE PROCESS INVOLVES SEVERAL **TWISTS AND TURNS**...

ESTIMATE!

STAB!

PUSH!

TURN AND LOOK!

CALCULATE!

DECIDE!

...SO WE'RE GOING TO **TAKE IT SLOWLY**.

SHSHSHSH.

THE PROCESS STARTS WHEN WE BUILD **ONE ESTIMATED SAMPLING DISTRIBUTION** WITH **ONE SAMPLE...**

REMEMBER, THAT THING IS LIKE A PORTRAIT OF **HOW OTHER SAMPLE AVERAGES WOULD PILE UP...**

...**IF** THEY WERE CENTERED ABOVE THE AVERAGE IN OUR ONE SAMPLE.

...AND **SLIDE** IT OVER TO **ANOTHER LOCATION THAT WE'RE CURIOUS ABOUT.**

WE USUALLY HAVE A **PARTICULAR LOCATION** IN MIND...

...BUT WE'LL GET TO THAT IN THE NEXT CHAPTER.

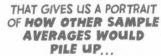

THAT GIVES US A PORTRAIT OF **HOW OTHER SAMPLE AVERAGES WOULD PILE UP...**

...**IF THEY WERE CENTERED AT THIS LOCATION!**

NEXT, WE **LOOK BACK** AT THE AVERAGE WE FOUND IN OUR ONE SAMPLE AND ASK:

HMM.

IF THE POPULATION AVERAGE **WERE REALLY RIGHT HERE...**

... **HOW LIKELY IS IT** THAT WE'D RANDOMLY GRAB A SAMPLE LIKE OURS?

WE CAN USE WHAT WE'VE LEARNED ABOUT **STATISTICAL CONFIDENCE** TO CALCULATE AN ANSWER!

IF OUR GUESS IS CORRECT, AND THE REAL POPULATION AVERAGE **IS RIGHT HERE...**

...THEN **IN THE LONG RUN**, WE'D EXPECT ANY ONE SAMPLE...

...TO HAVE AN AVERAGE SOMEWHERE **IN THE CLUMP UNDER THE HUMP.**

IT'S THE **GREAT DISCOVERY!**

YAY!

HOWEVER, **IF** OUR ONE SAMPLE HAS AN AVERAGE VALUE **WAY OUT HERE...**

LOTS SHORTER

...OR **WAY OUT HERE...**

LOTS LONGER

...WE HAVE REASON TO SUSPECT THAT OUR GUESS **MIGHT BE WRONG.**

I THINK **THIS** COULD BE IT, **REALLY!**

THEN **WHY** IS OUR OBSERVED AVERAGE WAY OVER HERE, EH?

BUT LET'S BE MORE PRECISE ABOUT IT:

IN THE LONG RUN, WE EXPECT **95%** OF ALL SAMPLE AVERAGES TO PILE UP **WITHIN TWO STANDARD DEVIATIONS OF THE REAL POPULATION AVERAGE...**

...SO THE **PROBABILITY** THAT WE'D RANDOMLY GRAB A SAMPLE AVERAGE WAY **OUT HERE...**

...OR **OUT HERE...**

...IS ONLY **ABOUT 5%.**

YOU **MIGHT** RANDOMLY GRAB A SAMPLE WAY OUT HERE...

...BUT IT'S **NOT VERY PROBABLE.**

IN PRACTICE, WE COMPARE OUR SAMPLE AND OUR GUESS BY CALCULATING SOMETHING CALLED A **PROBABILITY VALUE**＊ (OR "P-VALUE")...

...AND IF IT'S **LESS THAN 5%,** WE SUSPECT OUR GUESS **MIGHT BE WRONG.**

IF **THAT'S** THE REAL POPULATION AVERAGE...

...THE **PROBABILITY** WE'D GRAB A SAMPLE WAY OUT HERE IS ONLY ABOUT 4%.

SORRY, THAT'S A LITTLE **TOO UNLIKELY** FOR MY TASTE.

WE ALWAYS **FINISH** HYPOTHESIS TESTING BY **MAKING A FORMAL DECISION.**

IF OUR SAMPLE AND OUR GUESS ARE **FAIRLY CLOSE TOGETHER...**

WE GOT A P-VALUE OF **5% OR MORE** WHEN WE COMPARED THEM...

...WHICH MEANS OUR SAMPLE AVERAGE FITS **INSIDE THE 95% CLUMP UNDER THE HUMP.**

...WE HAVE TO **CONCLUDE** THAT OUR GUESS **MIGHT BE CORRECT.**

IT'S PRETTY LIKELY THAT WE'D RANDOMLY GRAB A SAMPLE LIKE OURS...

...FROM A POPULATION **CENTERED RIGHT THERE.**

HOWEVER, IF OUR SAMPLE AND OUR GUESS ARE **FAR APART**...

WE GOT A P-VALUE OF **LESS THAN 5%** WHEN WE COMPARE THEM...

...WHICH MEANS OUR SAMPLE AVERAGE IS **OUT IN THE TAILS!**

...WE CAN **CHOOSE TO REJECT IT.**

IT'S **INCREDIBLY UNLIKELY** THAT WE'D RANDOMLY GRAB A SAMPLE LIKE OURS...

...FROM A POPULATION **CENTERED RIGHT THERE.**

SO I'M THINKING THE REAL POPULATION **ISN'T ACTUALLY CENTERED RIGHT THERE.**

BUT THOSE ARE THE **ONLY TWO OPTIONS.**

UNFORTUNATELY, **NEITHER** OPTION IS **COMPLETELY SATISFYING.**

BLECH.

BECAUSE **NO MATTER WHICH FORMAL CONCLUSION WE REACH...**

IS **THIS** IT?

MAYBE SO!

OUR DATA **FIT PRETTY WELL** WITH THAT VALUE!

IS **THAT** IT?

IF THAT'S IT, WE **PROBABLY** WOULDN'T SEE DATA LIKE OURS.

SO I'M THINKING **THAT'S NOT IT.**

...WE **MIGHT** BE WRONG!*

ARGH!!

WHY CAN'T YOU GIVE ME A STRAIGHT ANSWER?

SORRY, BUT OUR RANDOM SAMPLE **COULD BE A DUD!**

* THIS CAN FEEL ANNOYING, BUT IT'S INCREDIBLY IMPORTANT. FOR MORE, SEE PAGE 222.

REMEMBER, HYPOTHESIS TESTING IS ALL BASED ON **ONE ESTIMATE** BUILT WITH **ONE SAMPLE**...

WE JUST **PUSH THE THING AROUND A BIT**.

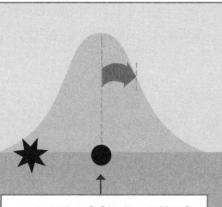

...AND IF OUR SAMPLE HAPPENS TO BE **MISLEADING**...

...ALL OUR CONCLUSIONS WILL **ALSO BE MISLEADING**.

IF WE RANDOMLY GRABBED **AN ABNORMAL SAMPLE**...

...**OUR P-VALUE** WON'T MEAN SQUAT!

LUCKILY, EVEN THOUGH **COMPLETE CERTAINTY** IS ALWAYS OUT OF REACH...

NO AMOUNT OF **TRAINING** CAN MAKE YOU CORRECT **100%** OF THE TIME.

...THERE **ARE** TIMES WHEN HYPOTHESIS TESTING CAN BE **EXTREMELY USEFUL**.

AS WE'LL SEE IN THE NEXT CHAPTER, THE **TWISTS AND TURNS** OF HYPOTHESIS TESTING...

ESTIMATE!

STAB!

PUSH!

TURN AND LOOK!

CALCULATE P-VALUE!

DECIDE!

...WORK BEST WHEN WE'RE WRESTLING WITH ONE **VERY PARTICULAR** KIND OF QUESTION:

IS THIS EVIDENCE **STRONG ENOUGH**...

...TO MAKE US **DOUBT** THIS HYPOTHESIS?

CHAPTER 13
SMACKDOWN

NOW THAT WE'VE LEARNED THE **FORMAL STEPS OF HYPOTHESIS TESTING...**

...LET'S SEE HOW THEY **ACTUALLY GET USED.**

ESTIMATE! **PUSH!**

CALCULATE! **DECIDE!**

GREAT, NOW GET OUT THERE AND **KICK SOME BUTT!**

IN **PRACTICE,** WE USE THESE STEPS...

...TO PIT ONE IDEA...

I'M NEW AND **DIFFERENT!**

...AGAINST ANOTHER.

MY NAME IS **NULL...**

...I'M **DULL.**

IT'S LIKE A **WRESTLING** **MATCH**...

...BUT IT HAS **LOPSIDED RULES**.

THE DULL IDEA ALWAYS WINS...

...UNLESS OUR EVIDENCE SEEMS **STRONG ENOUGH** TO **DEFEAT IT!**

TO SEE HOW IT WORKS, LET'S LOOK AT SOME **STORIES**.

DR. HAPPY GRABS
A **RANDOM SAMPLE**...

PUT ON THIS
BLINDFOLD, HON...

...AND GRAB
**80 SAMPLE
VIALS.**

...PLOTS IT OUT...

...AND CALCULATES THAT HER **SAMPLE AVERAGE IS
0.14 GRAMS.**

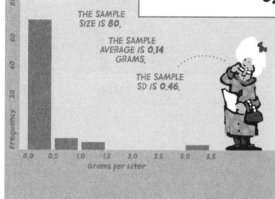

THE SAMPLE
SIZE IS 80,

THE SAMPLE
AVERAGE IS 0.14
GRAMS,

THE SAMPLE
SD IS 0.46.

THAT'S **NOT
ENOUGH** PURE EVIL
IN EVERY VIAL!

THAT OLD
MACHINE **MUST
BE BROKEN!**

BUT HERE'S THE **PROBLEM:**

DR. HAPPY **REALLY WANTS**
TO **BUY A NEW MACHINE**...

OH, LET'S JUST
**JUNK THE
BUGGER!**

I SO WANT TO
BUY THE NEW
XT-4300!

...BUT HER ACCOUNTANT WON'T LET HER
UNLESS THEY CAN **BE CONFIDENT** THAT
THE MACHINE IS **REALLY BROKEN.**

MAYBE THE OLD MACHINE
**ACTUALLY STILL
WORKS JUST FINE**...

...AND OUR ONE SAMPLE
IS LIGHT ON PURE EVIL
JUST BY CHANCE!

TO **MEASURE HER CONFIDENCE,**
DR. HAPPY CAN USE A **HYPOTHESIS TEST.**

179

DR. HAPPY'S HYPOTHESIS TEST PITS **THESE TWO IDEAS AGAINST EACH OTHER:**

EITHER THAT SAD, OLD, TIRED MACHINE **IS BUSTED**...

...OR IT **ISN'T**.

ONE IDEA IS **EXCITING**...

...AND ONE IS **DULL**.

I GET TO **BUY A NEW ONE!**

LET'S NOT **RUSH INTO ANYTHING**, MINERVA.

EACH IDEA COMES WITH A **DIFFERENT EXPLANATION** FOR WHY WE GOT THE DATA WE DID...

THIS SAMPLE AVERAGE IS LIGHT...

...BECAUSE THE MACHINE **JUST CAN'T INJECT ENOUGH PURE EVIL INTO THE VIALS ANYMORE!**

ANY SINGLE SAMPLE WILL PROBABLY BE A BIT **LIGHT** OR **HEAVY** JUST BY CHANCE.

WHAT IF THIS ONE IS LIGHTISH **JUST BECAUSE OF RANDOM VARIATION?**

...AND DR. HAPPY'S TASK IS TO LOOK AT THE **DULL** EXPLANATION AND **SEE IF SHE CAN REJECT IT!**

SO DR. HAPPY USES HER DATA TO **BUILD AN ESTIMATED SAMPLING DISTRIBUTION.**

I BUILT THIS WITH 80 RANDOM VIALS.

THEN SHE SLIDES IT TO THE LOCATION PREDICTED BY HER **DULL HYPOTHESIS...**

I'LL PUSH IT OVER TO WHERE IT'D BE CENTERED **IF THE MACHINE ISN'T BROKEN.**

...AND CALCULATES A **P-VALUE...**

IF THE POPULATION WERE STILL CENTERED AT 0.25...

...THERE'S ONLY A 3% CHANCE...

...THAT I'D RANDOMLY GRAB A SAMPLE AVERAGE LIKE THE ONE I GRABBED!

...THEN PAUSES TO THINK ABOUT WHAT HER P-VALUE **ACTUALLY MEANS.**

A P-VALUE OF **LESS THAN 5%...**

...MEANS THAT **IN THE LONG RUN** WE'D EXPECT TO RANDOMLY GRAB...

...FEWER THAN 1 OUT OF **20** SAMPLE AVERAGES LIKE THE ONE I GRABBED...

...IF THE MACHINE WAS STILL AVERAGING 0.25 g.

FINALLY, SHE **CONFIDENTLY REJECTS** THE DULL EXPLANATION...

...IN FAVOR OF **THE EXCITING ONE.**

SUGAR, THAT PROBABILITY IS SO **EENSY WEENSY...**

...THAT I **DON'T THINK** WE GOT A LIGHTISH AVERAGE **JUST BY CHANCE.**

I GET TO **BUY A NEW MACHINE!**

IN THIS STORY, **DR. HAPPY'S P-VALUE** HELPED HER **MAKE A CONFIDENT DECISION.**

A P-VALUE OF **0.03** MEANS THAT...

...I CAN BE **97% CONFIDENT** ABOUT IT!

BUT REMEMBER, IN STATISTICS, **CONFIDENCE ALWAYS HAS A FLIP SIDE.**

A P-VALUE OF **0.03 ALSO** MEANS THAT...

...IN THE LONG RUN, WE EXPECT TO GRAB SAMPLES AS MISLEADING AS THIS ONE 3% OF THE TIME...

...AND **MAYBE** WE JUST DID.

SO EVEN THOUGH THE EVIDENCE **SEEMS TO SUPPORT** DR. HAPPY'S DECISION...

...SHE **COULD** BE WRONG.

I'M BUYING A NEW ONE AND **YOU CAN'T STOP ME!**

THAT RUSTY OLD HUNK O' CRAP NEEDS TO BE **PUT OUT OF ITS MISERY!**

I'M JUST SAYIN' IT **MIGHT** STILL BE WORKING JUST FINE.

WE CAN **NEVER BE CERTAIN ABOUT RANDOM SAMPLES.**

FOR BETTER OR FOR WORSE, HYPOTHESIS TESTS **ALWAYS** END THIS WAY.

MY DECISION **MIGHT BE WRONG!**

FORTUNATELY, IN THE LONG RUN, IT **PROBABLY ISN'T.**

HAND ME MY **CREDIT CARD**, HON. I'M ORDERING THE NEW MACHINE!

JEEZ, KEEP YOUR **KNICKERS ON.**

NEVERTHELESS, IN THE END, DR. HAPPY **GOT WHAT SHE WANTED** FROM HER HYPOTHESIS TEST...

I CAN **DOUBT THE DULL ONE!**

I GET TO BUY THE XT-4300!

I HAVE THE **BEST JOB** IN THE WORLD!

...BUT THAT'S **NOT ALWAYS** THE CASE.

TO SEE HOW, LET'S TELL **ANOTHER STORY**.

IMAGINE THAT **CRAZY BILLY** HAS BEEN **POURING WORM STEROIDS** INTO HIS SWAMP...

THIS STUFF IS **GUARANTEED** TO MAKE YOUR WORMS GROW LONGER!

BUT IT'S **NOT CHEAP!**

worm 'Roids

...AND HE WANTS TO **FIND OUT IF THEY'RE WORKING.**

SO HE GRABS A **RANDOM SAMPLE**...

...28

...29

...30

...PLOTS IT OUT...

THE SAMPLE SIZE IS 30.

THE SAMPLE AVERAGE IS 4.19 INCHES.

THE SAMPLE SD IS 0.34.

Frequency

0 1 2 3 4 5 6 7 8
Length (inches)

...AND CALCULATES THAT HIS **SAMPLE AVERAGE IS 4.19 INCHES.**

HMMM, THAT'S **ENCOURAGING.**

BEING BILLY, HE **KNOWS** THAT THE SWAMP WORM POPULATION AVERAGE **USED TO BE 4 INCHES...**

I REMEMBER EVERY CAN I'VE EVER SOLD...

...AND I'VE SOLD A **GAZILLION** OF THEM!

...AND HE'S **HOPING** TO PROVE **THAT IT HAS GOTTEN LONGER.**

IF IT'S LONGER...

...THESE 'ROIDS **ARE WORKING!**

IF IT'S **NOT** LONGER...

...I'VE BEEN **HOODWINKED.**

HERE'S THE **PROBLEM:**

HE'S GOT A SAMPLE AVERAGE IN HIS HAND THAT'S LONGER **THAN THE OLD AVERAGE...**

THIS SAMPLE MAKES ME THINK **MAYBE THE 'ROIDS ARE WORKING!**

...BUT IT **MIGHT** BE LONGER **JUST BY CHANCE!**

MAYBE I JUST **RANDOMLY** GRABBED 30 **ABNORMALLY LONG WORMS...**

...FROM A POPULATION THAT **HASN'T** CHANGED!

TO DECIDE WHETHER HE THINKS THAT'S THE CASE, BILLY CAN USE A **HYPOTHESIS TEST.**

CRAZY BILLY'S HYPOTHESIS TEST PITS
THESE TWO IDEAS AGAINST EACH OTHER:

EITHER MY **WORM 'ROID REGIMEN IS WORKING...**

...OR IT **ISN'T.**

ONE IDEA IS **NEW AND DIFFERENT...**

...AND ONE IS **RATHER DULL.**

THE POPULATION AVERAGE HAS **ACTUALLY CHANGED!**

THE POPULATION IS **THE SAME AS IT EVER WAS.**

EACH IDEA COMES WITH A **DIFFERENT EXPLANATION** FOR WHY WE GOT THE DATA WE DID...

MY NEW SAMPLE IS LONGER **BECAUSE THE AVERAGE IN THE SWAMP IS LONGER!**

MY NEW SAMPLE IS LONGER **BECAUSE I GRABBED LONGISH WORMS JUST BY CHANCE.**

...AND BILLY'S TASK IS TO LOOK AT THE **DULL** EXPLANATION AND **SEE IF HE CAN REJECT IT!**

SO CRAZY BILLY USES HIS DATA TO **BUILD AN ESTIMATED SAMPLING DISTRIBUTION.**

I BUILT THIS WITH **30** WORMS.

THEN HE SLIDES IT TO THE LOCATION PREDICTED BY HIS **DULL HYPOTHESIS...**

I'LL PUSH IT OVER TO WHAT I KNOW WAS THE **OLD POPULATION AVERAGE.**

...AND CALCULATES A **P-VALUE.**

IF THE POPULATION WERE STILL CENTERED AT **4**...

...THERE'S A **28%** CHANCE...

...THAT I'D RANDOMLY GRAB A SAMPLE AVERAGE LIKE THE ONE I GRABBED.

IN THIS CASE WE ONLY LOOK AT **ONE** SHADED END BECAUSE WE'RE FOCUSED ONLY ON WHETHER THE POPULATION AVERAGE HAS GOTTEN **LONGER.**

BUT WHEN HE REALIZES WHAT HIS P-VALUE **ACTUALLY MEANS...**

A P-VALUE OF **28%**...

...MEANS THAT **IN THE LONG RUN** WE'D EXPECT TO SEE DATA LIKE MINE ABOUT **3** OUT OF **10** TIMES...

...IF THE REAL POPULATION AVERAGE WAS STILL **4**.

&%@#$!

...HE SADLY CONCLUDES THAT HE **CAN'T BE CONFIDENT** THAT THE 'ROIDS ARE WORKING!

IT SEEMS PERFECTLY POSSIBLE THAT WE GOT A LONGISH AVERAGE **JUST BY CHANCE.**

OF COURSE, BILLY'S CONCLUSION...
...DOESN'T MEAN HIS 'ROIDS AREN'T WORKING.
BUMMER!
I CAN'T BE CONFIDENT THE 'ROIDS ARE WORKING.
BUCK UP,
THEY MIGHT ACTUALLY BE WORKING!
YOUR SAMPLE AVERAGE IS LONGER THAN THE OLD AVERAGE, AFTER ALL.
IT JUST MEANS HIS EVIDENCE ISN'T STRONG ENOUGH TO SUPPORT THE HYPOTHESIS HE WANTED.
YOU COULD HAVE GOTTEN THESE RESULTS JUST BY CHANCE!
IT'S NOT A TERRIBLY SATISFYING CONCLUSION TO HIS TEST...
NOTHING REALLY EXCITING GOING ON HERE.
WE'RE BACK WHERE WE STARTED.
...BUT IT'S AN IMPORTANT ONE.

WE'VE NOW SEEN **TWO DIFFERENT HYPOTHESIS TESTS...**

I WANTED TO SEE EVIDENCE THAT MY MACHINE WAS BROKEN.

I WANTED TO SEE EVIDENCE OF LONGER WORMS.

... WITH **TWO DIFFERENT OUTCOMES.**

I CAN FEEL **CONFIDENT** ABOUT MY EVIDENCE...

...I CAN'T.

BOTH STORIES PITTED **AN ATTRACTIVE NEW IDEA...**

...AGAINST A **TIRED, OLD, DULL ONE.**

I'M **SEXY** IN SPANDEX!

MEH.

IN BOTH STORIES WE **WANTED THE NEW IDEA TO BE TRUE...**

YOU KNOW **YOU WANT ME TO WIN!**

... BUT **GAVE THE DULL IDEA THE BENEFIT OF THE DOUBT.**

BUT UNLESS YOU HAVE ENOUGH **EVIDENCE TO REJECT ME...**

...I WIN.

THERE'S A **GOOD REASON** WE DO HYPOTHESIS TESTING LIKE THIS.

CHAPTER 14
FLYING PIGS, DROOLING ALIENS, AND FIRECRACKERS

LOOKS LIKE **STORMY WEATHER.**

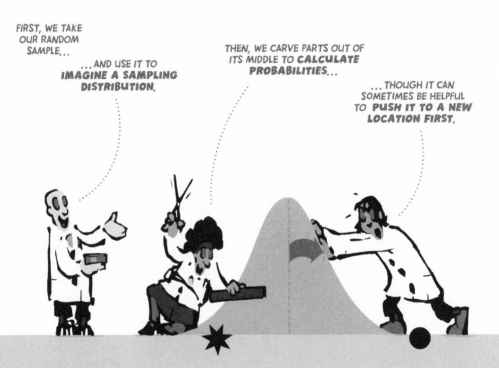

NOW THAT WE'VE COVERED THE **FUNDAMENTALS**...

CONGRATULATIONS!

IF YOU UNDERSTAND HOW **THAT STUFF** WORKS...

...YOU **UNDERSTAND STATISTICS!**

THE REST IS MOSTLY **DETAILS.**

...WE'RE GOING TO SPEND THIS CHAPTER LOOKING AT A FEW **STORIES**...

...TO GIVE YOU A SENSE OF WHAT LIES AHEAD **IF YOU WANT TO LEARN MORE.**

THE FORECAST CALLS FOR **FLYING PIGS, DROOLING ALIENS,** AND **FIRECRACKERS!**

WE'RE GOING TO WANT **THIS.**

THUS FAR, WE'VE FOCUSED ON HOW OUR **BASIC STEPS** WORK **IN IDEAL CONDITIONS:**

WE'VE LEARNED TO HUNT FOR **ONE POPULATION AVERAGE...**

...USING ONE **LARGE SAMPLE...**

...OF TRULY **RANDOM MEASUREMENTS.**

THE **WATER** IS CLEAR...

THE **SUN** IS SHINING...

...STATISTICS IS **EASY!**

IN PRACTICE, HOWEVER, CONDITIONS ARE OFTEN **MORE COMPLEX...**

WHAT IF YOU **CAN'T** GET A LARGE SAMPLE SIZE?!

WHAT IF YOU CAN'T GET MEASUREMENTS THAT ARE **REALLY RANDOM?!**

WHAT IF YOU WANT TO KNOW ABOUT **SOMETHING THAT'S NOT AN AVERAGE?**

...AND MUCH OF **ADVANCED STATISTICS** INVOLVES GRAPPLING WITH THESE COMPLEXITIES.

THE **GOOD NEWS** IS THAT NO MATTER HOW **COMPLICATED** THINGS GET...

...AND THEY **DO** GET **COMPLICATED**...

MY KINGDOM FOR A **P-VALUE!**

I'M NOT FEELING VERY **CONFIDENT!**

...WE CAN STILL RELY ON THE **BASIC STEPS** WE'VE COVERED IN THIS BOOK...

... WE JUST HAVE TO LEARN HOW TO **MODIFY** THEM.

THE **DETAILS** CHANGE...

...BUT THE **BASIC STEPS** REMAIN THE SAME.

FLYING PIGS!

IN OUR FIRST STORY, LET'S IMAGINE THAT **SPOTTED** FLYING PIGS ARE **FASTER THAN STRIPED** FLYING PIGS...

EAT MY DUST, CUPCAKE!

...BUT THEY'RE ALSO **LOTS** MORE **EXPENSIVE**...

...AND **REVOLTING!**

AND THAT'S WHY SAM NEEDLEHOUSE WANTS TO KNOW: **HOW MUCH FASTER ARE THEY?**

I'M STARTING A **PIG-O-GRAM** BUSINESS.

IS IT BETTER TO **INVEST IN SPOTTED PIGS**,...

...OR SHOULD I GET **STRIPED PIGS** INSTEAD, AND USE THE MONEY I SAVE TO **BUY CUTE COSTUMES FOR THEM?**

DO YOU WANNA INVEST IN **SPEED** OR IN **STYLE?**

TO MAKE THAT DECISION, YOU WANT TO HAVE A SENSE OF **HOW MUCH FASTER SPOTTED PIGS ARE.**

IN **STATISTICAL TERMS,** HERE'S THE QUESTION:

HOW DO WE CONSTRUCT A **CONFIDENCE INTERVAL**...

...THAT TELLS US ABOUT THE **DIFFERENCE** BETWEEN **TWO SEPARATE POPULATION AVERAGES?**

LET'S GRAB **TWO SETS** OF RANDOM SAMPLE DATA TO FIND OUT.

IN THIS CASE, WE CAN USE DATA FROM **40 RANDOM SPOTTED PIGS**...

...AND **40 RANDOM STRIPED PIGS**...

OUR SAMPLE AVERAGE IS **59.7 MPH**.

OUR SAMPLE SD IS **4.6 MPH**.

OUR SAMPLE AVERAGE IS **44.2 MPH**.

OUR SAMPLE SD IS **4.7 MPH**.

... TO BUILD AN **ESTIMATED SAMPLING DISTRIBUTION**...

... THAT'S **SLIGHTLY DIFFERENT** FROM THE ONE WE'RE USED TO.

WE CENTER IT AT **THE DIFFERENCE** BETWEEN OUR SAMPLE AVERAGES,...

...AND WE CALCULATE THE **STANDARD DEVIATION** A BIT DIFFERENTLY,...

...BUT THE DISTRIBUTION IS **STILL NORMAL-SHAPED!**

THAT'S THEIR BEST GUESS AT HOW **DIFFERENCES BETWEEN SAMPLE AVERAGES** WOULD LOOK IF THEY RANDOMLY GRABBED A GAZILLION SAMPLES FROM EACH POPULATION.

15.5

13.5   15.5   17.5

IT'S SLIGHTLY DIFFERENT, BUT WE CAN STILL **CARVE IT UP**...*

BECAUSE IT'S NORMAL, WE CAN CHOP OFF THE TAILS **2 SDs** AWAY FROM THE CENTER AND SAY:

WE'RE **95%** CONFIDENT,...

...THAT ON AVERAGE SPOTTED PIGS ARE BETWEEN **13.5** AND **17.5** MPH FASTER THAN STRIPED PIGS.

...AND USE IT TO MAKE A **CONFIDENT DECISION**.

I'LL **GET SPOTTED PIGS!**

IF THEY'RE **THAT MUCH FASTER**...

...IN THE LONG RUN, THE EXTRA MONEY WILL BE **WORTH IT!**

* SEE **PAGE 223** FOR THE FORMULA.

DROOLING ALIENS!

THE **MAN-EATING BUG ALIENS** FROM THE PLANET E8M-286 HAVE **ACIDIC SALIVA**...

I RECOMMEND **NOT** USING IT AS A FACIAL MOISTURIZER!

...AND WE WANT TO KNOW, **HOW ACIDIC IS IT?**

WHAT'S THEIR **AVERAGE SALIVA pH?**

IS IT GOING TO **BURN THROUGH OUR BODY ARMOR?**

UNFORTUNATELY, WE CAN GRAB AND MEASURE ONLY **A FEW OF THEM**...

...WHICH **ISN'T ENOUGH** TO MAKE OUR INFERENCE TOOLS WORK THE WAY WE'VE LEARNED SO FAR.

WE COULD GET DATA FROM ONLY **10 RANDOM ALIENS**...

...BEFORE THEY **ATE** OUR LAST DATA COLLECTION OFFICER.

LESS THAN 30 IS A **SMALL** SAMPLE SIZE...

...WE'LL HAVE TO USE **DIFFERENT MATH!**

LUCKILY, THERE'S A **CLEVER STRATEGY** WE CAN USE WHEN WE'RE STUCK WITH A **SMALL SAMPLE SIZE**...

WE'D **ALWAYS** LIKE TO GRAB MORE DATA, **BUT WE CAN'T!**

THERE ARE ONLY **10** OF US.

OUR SAMPLE AVERAGE IS **2.38.**

OUR SAMPLE SD IS **0.48.**

More Acidic ← Saliva pH → Less Acidic

Frequency

...BUT IT BEGINS WITH ONE **WHOPPING ASSUMPTION.**

IF WE'RE WILLING TO **ASSUME** THAT THE **OVERALL POPULATION IS NORMAL-SHAPED...**

SEEMS LIKE A LOT OF **NATURAL PHENOMENA** ARE NORMALLY DISTRIBUTED, SO MAYBE THIS ONE IS, TOO.

... WHICH CAN BE A **DUBIOUS ASSUMPTION...**

IN THIS CASE, IT'S HARD TO TELL WITH JUST **10 ALIENS.**

... WE CAN USE OUR SAMPLE DATA TO BUILD AN **ESTIMATED SAMPLING DISTRIBUTION...**

... THAT'S GOT A SLIGHTLY **FATTER SHAPE** THAN THE ONE WE'RE USED TO.

WE CENTER IT AT OUR SAMPLE AVERAGE...

2.38

... AND WE CALCULATE THE **STANDARD DEVIATION** THE SAME OLD WAY...

IT LOOKS SIMILAR, BUT IT HAS **MORE PROBABILITY OUT IN THE TAILS...**

... BUT IT'S **NOT NORMAL-SHAPED!**

... IT'S CALLED A **T-DISTRIBUTION.**

0.15   0.15   0.15   0.15

2.04   2.38   2.72

IT'S FATTER, BUT WE CAN STILL USE IT TO **CALCULATE OUR CONFIDENCE...**＊

FOR A 95% CONFIDENCE INTERVAL IN THIS KIND OF T-DISTRIBUTION, WE COUNT OUTWARD **2.26 SDS** INSTEAD OF JUST **2.**

WHICH MEANS WE'RE **95% CONFIDENT...**

... THAT THEIR AVERAGE SALIVA pH IS BETWEEN **2.04** AND **2.72.**

... WE JUST HAVE TO BE **EXTRA CAREFUL ABOUT OUR CONCLUSIONS.**

IF OUR ASSUMPTION ABOUT A NORMAL-SHAPED POPULATION IS **WRONG...**

WE MIGHT GET MELTED.

THAT'S SOMEWHERE BETWEEN **VINEGAR** AND **LEMON JUICE!**

＊ SEE **PAGE 223** FOR THE FORMULA.

UNFORTUNATELY, LITTLE SUZIE BICKER WANTS TO **HARM THE NEIGHBOR'S CAT**...

HERE, KITTY KITTY KITTY.

...WITH **FIRECRACKERS!**

SSSSSSSSSSSSS,

SHE PREFERS A BRAND CALLED **DINGALINGS**...

...BECAUSE THE PACKAGE CLAIMS THEY HAVE AN **AVERAGE FUSE TIME OF FIVE SECONDS**...

POW!

MEOW!

THAT'S THE **PERFECT** LENGTH OF TIME!

...BUT SHE'S CONCERNED ABOUT **HOW DEPENDABLE THEY ARE**.

THE AVERAGE MAY BE FIVE SECONDS...

BUT SOME EXPLODE **QUICKER**, *AND SOME* **SLOWER**.

I DON'T LIKE IT WHEN THEY **EXPLODE IN MY HAND**...

...OR **TAKE SO LONG THAT THE CAT HAS TIME TO RUN AWAY!**

SSSSSSSSSSSSSSSSSSSSSSSSSSS,

IN **STATISTICAL TERMS,** SUZIE'S QUESTION IS ABOUT **VARIABILITY**...

...AND WE CAN TACKLE IT BY USING ONE SAMPLE...

I WANNA KNOW, WHEN I LIGHT A DINGALING, **CAN I EXPECT IT TO EXPLODE IN CLOSE TO THE AVERAGE TIME?**

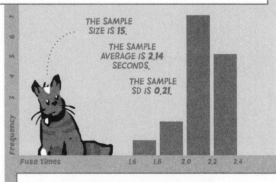

THE SAMPLE SIZE IS **15**.

THE SAMPLE AVERAGE IS **2.14** SECONDS.

THE SAMPLE SD IS **0.21**.

...TO MAKE A GUESS ABOUT THE **SPREAD** OF THE OVERALL POPULATION.

IN THIS CASE, WE WANT TO DO INFERENCE ON A **STANDARD DEVIATION** INSTEAD OF AN AVERAGE...

... AND THE WHOLE PROCESS REQUIRES **VERY DIFFERENT MATH.**

FIRE IN THE HOLE!

SSSSSSSSSSSSSSSSSSSS.

STEP BACK! WE'RE **TRAINED PROFESSIONALS.**

DON'T TRY THIS AT HOME!

NEVERTHELESS, WE STILL USE THE SAME **BASIC STEPS:**

WE BUILD A SLIGHTLY DIFFERENT KIND OF **ESTIMATED SAMPLING DISTRIBUTION...** *

... AND **CARVE PROBABILITIES** OUT OF IT...

THIS IS OUR BEST GUESS AT HOW DINGALING **SAMPLE STANDARD DEVIATIONS** WOULD LOOK IF WE RANDOMLY GRABBED A GAZILLION OF THEM!

IT'S TOTALLY NOT NORMAL, IT'S **SKEWED!**

SO HOW DO WE KNOW WHERE TO CARVE A **95%** CONFIDENCE INTERVAL?

OR HOW TO CALCULATE A P-VALUE?

YOU HAVE TO **ASK YOUR COMPUTER.**

0.10   **0.16**   **0.28**   **0.34**   **0.40**

... TO REACH A **FAMILIAR CONCLUSION.**

BASED ON OUR SAMPLE, WE'RE **95%** CONFIDENT...

... THAT THE STANDARD DEVIATION OF THE DINGALING POPULATION IS BETWEEN **0.16** AND **0.34** SECONDS...

... AND IF THAT'S TRUE, YOU CAN EXPECT **MOST DINGALINGS** TO EXPLODE BETWEEN 1.5 AND 2.4 SECONDS!

HEH HEH

SSSSSSSSS.

* SEE PAGE 224 FOR SOME TECHNICAL DETAILS.

AS THESE STORIES SUGGEST, WE DEPEND ON A **DEEP BAG OF TRICKS**...

... WHEN WE GRAPPLE WITH **ADVANCED STATISTICS QUESTIONS.**

WHAT IF WE WANT TO COMPARE **DROOLING ALIENS** TO **FIRECRACKERS?**

JUST A SEC.

AND IN TRUTH, THE BAG IS **NEARLY BOTTOMLESS.**

FOR EXAMPLE, IF WE HAVE TO CONTEND WITH DATA THAT'S **CORRELATED*** IN SOME WAY...

* SEE **PAGE 224** FOR A MORE TECHNICAL EXPLANATION.

WE WANT TO KNOW THE AVERAGE TEMPERATURE OF ALL THE **GECKOS** IN THIS RAIN FOREST.

BUT WE CAN'T GET **TRULY RANDOM DATA** BECAUSE THE GECKOS **IN THE SUN** ARE **WARMER**...

...THAN THE GECKOS **IN THE SHADE.**

WHICH MEANS OUR TEMPERATURES WILL BE **CORRELATED WITH ONE ANOTHER**...

...AND **SO WILL OURS.**

... WE HAVE **CERTAIN TRICKS** WE CAN USE.

IF WE STRAP A **CORRELATION STRUCTURE** AROUND THE GECKOS...

...WE CAN STILL USE THEM TO **ESTIMATE A SAMPLING DISTRIBUTION!**

HOW DOES YOUR **RATE OF SHRINKAGE**...

...DEPEND ON **HOW MUCH SHRINKING MEDICINE YOU DRINK?**

...WE HAVE **ENTIRELY DIFFERENT TRICKS** AT OUR DISPOSAL.

WE CAN DO **REGRESSION ANALYSIS**...

...WHICH INVOLVES DRAWING A **LINE BETWEEN** TWO QUALITIES ON THE SAME GRAPH...

...AND ESTIMATING A SAMPLING DISTRIBUTION **WITH THE SLOPE OF THAT LINE!**

Shrinkage (inches)

Dose (grams)

THE POINT IS, EVEN THOUGH **ADVANCED STATISTICS** IS **CRAMMED FULL** OF TRICKS AND **TWEAKS**...*

DON'T FORGET **ANOVA**...

...AND HOW TO **DO INFERENCE ON PROPORTIONS**...

...AND HOW TO **PREDICT THE FUTURE!**

...THE **BASIC STEPS** OF STATISTICAL INFERENCE **REMAIN THE SAME!**

* SEE **PAGES 224–225** FOR A MORE THOROUGH RUNDOWN.

SO KEEP THIS IN MIND IF YOU GO ON TO **LEARN MORE STATISTICS**:

THE DETAILS CAN SEEM **OVERWHELMING** AT FIRST...

IF YOU WANT TO KNOW HOW TO **PREDICT THE WEATHER**...

...HERE'S **A WHOLE SEPARATE BAG!**

...BUT AT THEIR HEART, **ALL STATISTICS PROBLEMS ARE SIMILAR.**

THEY **LOOK LIKE THIS**:

HOW DO WE MAKE JUDGMENTS ABOUT **POPULATIONS**...

...WHEN WE ONLY HAVE ACCESS TO **SAMPLES**?

AND WE TACKLE THEM **LIKE THIS:**

WE USE OUR DATA TO ESTIMATE SOME KIND OF **SAMPLING DISTRIBUTION**...

...THEN WE **CARVE PROBABILITIES OUT OF IT**...

...THOUGH IT CAN SOMETIMES BE HELPFUL TO **PUSH IT TO A NEW LOCATION FIRST.**

CONCLUSION
THINKING LIKE A STATISTICIAN

OMMMMMM.

IN THIS BOOK WE'VE BEEN **FISHING**...

I WISH WE COULD CATCH **ALL** THE PIRANHAS...

...BUT WE CAN ONLY CATCH **SOME**.

...**GATHERING**...

YOUR **HELMET** IS TO KEEP YOU SAFE.

THIS WILL HELP YOU **AVOID BIAS**.

...AND **HUNTING**.

WE'LL **NEVER** BE ABLE TO ACTUALLY CATCH IT...

...BUT WE CAN MAKE A GUESS ABOUT THE **CLUSTER** THAT SURROUNDS IT.

AND ALL THE WHILE, WE'VE BEEN LEARNING **HOW STATISTICIANS THINK!**

WE DON'T KNOW **EVERYTHING**...

...BUT THAT DOESN'T MEAN WE KNOW **NOTHING**!

IN **PART ONE**, WE EXAMINED PILES OF SAMPLE DATA...

TELL US ABOUT YOUR **SHAPE**, **LOCATION**, AND **SPREAD**.

ARGH! WE'RE **SKEWED!**

HAR HAR.

...AND **INVESTIGATED THEM**.

BEWARE **LURKING VARIABLES!**

THEN, IN **PART TWO**, WE STUDIED **STATISTICAL INFERENCE**...

WHAT DO **THESE WORMS**...

...TELL US ABOUT **THOSE WORMS?**

IT'S ALL ABOUT CALCULATING **PROBABILITIES!**

...WHICH IS HOW WE USE **SAMPLES** TO **SEARCH FOR QUALITIES IN AN OVERALL POPULATION**.

MORE SPECIFICALLY, WE LEARNED HOW TO BUILD **ESTIMATED SAMPLING DISTRIBUTIONS**...

HERE'S ONE CAN.

HERE'S A **BLUEPRINT** OF HOW A **GAZILLION CANS** WILL **TEND TO CLUMP** IN THE LONG RUN.

IT'S THE **CENTRAL LIMIT THEOREM**, YAY!

Y'ALL ARE THE **CRAZY ONES!**

...AND HOW TO **PEER** INTO THEM...

YOO HOO, ARE YOU **IN** THERE?

...AND **CARVE THEM UP** TO **CALCULATE CONFIDENCE INTERVALS**...

BASED ON THIS RANDOM SAMPLE...

WE'RE **95% CONFIDENT**...

...THAT THEY **DON'T** HATE YOU!

...OR **PUSH THEM AROUND** TO PERFORM **HYPOTHESIS TESTS**.

UM, I'M PRETTY SURE THE POPULATION AVERAGE IS **RIGHT HERE**.

WHAT DO YOU MAKE OF **THIS**, THEN?

FINALLY, WE LEARNED HOW WE CAN **MODIFY** THESE BASIC STEPS...

... WHEN OUR QUESTIONS GET **MORE COMPLICATED**.

WHEN CIRCUMSTANCES CHANGE...

... WE USE **DIFFERENTLY SHAPED SAMPLING DISTRIBUTIONS!**

POW!

SINCE THE **BASIC STEPS** OF STATISTICAL INFERENCE WERE FIRST DISCOVERED...

SACRÉ BLEU! THAT IS ONE **BEAUTIFUL SHAPE!**

...THEY'VE BEEN **ADAPTED** BY ALL KINDS OF DIFFERENT PEOPLE...

INTERNATIONAL SPIES!

COSMOLOGISTS!

BEER BREWERS!

GENERALS!

...FOR USE IN ALL KINDS OF **DIFFERENT CIRCUMSTANCES.**

I'M 68% CONFIDENT THAT I UNDERSTAND **THIS SECRET CODE!**

I'M 95% CONFIDENT THAT **THE UNIVERSE** IS BETWEEN **12** AND **15** BILLION YEARS OLD.

I'M 3% CONFIDENT THAT **THIS BATCH** TASTES **GREAT**... ...**AND** IS LESS FILLING.

I'M 99.7% CONFIDENT THAT **THIS WAR** IS A TERRIBLE IDEA... ...BUT LET'S **DO IT ANYWAY!**

WHILE THIS IS A TESTAMENT TO THE INCREDIBLE **POWER** OF STATISTICS...

WE CAN USE STATISTICS **WHENEVER** WE HAVE **LIMITED INFORMATION**...

...AND WANT TO MAKE **CONFIDENT DECISIONS.**

...IT HAS CREATED A BIT OF A **CONUNDRUM.**

OVER THE YEARS, **STATISTICAL TERMINOLOGY** HAS **FLOURISHED**...

...AND **MULTIPLIED**...

THE **NORMAL** DISTRIBUTION...

...IS ALSO CALLED THE **GAUSSIAN** DISTRIBUTION...

...IS **ALSO** CALLED THE **Z-DISTRIBUTION**.

...UM.

...AND AS A RESULT, LEARNING HOW TO **TALK** LIKE A STATISTICIAN...

...CAN BE **ROUGH**...

OUCH, I THINK MY **EVENTS** MIGHT BE **DISJOINT**!

IS MY **DEGREE OF FREEDOM RESIDUAL**?

DO I WANT **SIGNIFICANCE** OR **POWER**?

IS THIS **ERROR** STANDARD?

...ESPECIALLY IF YOU MOVE ON TO MASTER THE MORE **ADVANCED TOOLS**.

REPEAT AFTER ME:

P-VALUE,

E-VALUE,

Z- OR **T-SCORES,**

CHI-SQUARED **G-TESTS,**

AND THERE'S EVEN **MORE**!

I'M NOT FEELING VERY **CONFIDENT** AT THE MOMENT.

HAVE YOU CONSIDERED A **NON-PARAMETRIC REGRESSION SIMULATION**?

KEEP **ONE EYE**
ON THE LONG
RUN...

...AND **ONE EYE**
ON THE SHORT
RUN...

...AT THE
SAME TIME.

ANYONE CAN
DO IT!

BUT IF YOU WANT TO LEARN
HOW THEY TALK...

...YOU CAN **START**
BY EXPLORING...

YOU'LL
WANT SOME
GLOVES...

...AND A
HELMET.

APPENDIX
THE MATH CAVE
ABANDON HOPE ALL YE WHO ENTER HERE...
...UNLESS YOU WANT TO LEARN TO TALK LIKE A STATISTICIAN...
...AND WRITE LIKE ONE, TOO!
$\bar{x} \pm 2\left(\frac{s}{\sqrt{n}}\right)$
$\bar{x} = \frac{1}{n}\sum_{i=1}^{n} x_i$
$s = \sqrt{\frac{1}{n-1}\sum_{i=1}^{n}(x_i - \bar{x})^2}$

RANDOM SAMPLING

FROM PP. 36-37

RANDOM SAMPLING IS ABSOLUTELY ESSENTIAL TO STATISTICAL INQUIRY.
THE KEY FEATURE OF A RANDOM SAMPLE IS THAT IT **DOES NOT DIFFER
SYSTEMATICALLY** FROM THE POPULATION IT COMES FROM.

TECHNICALLY, A **SAMPLE** IS A COLLECTION OF SEPARATE OBSERVATIONS
ABOUT A SPECIFIC **VARIABLE** (SEE BELOW). WE CALL IT A **RANDOM
SAMPLE** WHEN IT'S MADE UP OF RANDOMLY GATHERED OBSERVATIONS,
EACH OF WHICH IS **INDEPENDENT** OF ALL THE OTHERS.

WHEN WE TALK IN THIS BOOK ABOUT RANDOM SAMPLING, WE
SPECIFICALLY MEAN **SIMPLE RANDOM SAMPLING**. FORMALLY,
A **SIMPLE RANDOM SAMPLE (SRS)** OF SIZE n IS A COLLECTION
OF n OBSERVATIONS OBTAINED IN SUCH A WAY THAT **ALL POSSIBLE
SAMPLES** OF n OBSERVATIONS FROM THE POPULATION ARE **EQUALLY
LIKELY TO HAVE BEEN SELECTED**.

SOME OTHER **NON-RANDOM SAMPLING TECHNIQUES** SUCH AS
SYSTEMATIC SAMPLING AND STRATIFIED SAMPLING SOMETIMES ALSO
WORK, BUT WHATEVER SAMPLING STRATEGY WE END UP USING, WE MUST
BE CERTAIN THAT THE RESULTING SAMPLE IS **REPRESENTATIVE OF THE
POPULATION**. IF IT'S NOT, EVERYTHING THAT FOLLOWS IS WORTHLESS.

$X_1, X_2, X_3 \ldots X_n$

SO THAT X_1 IS THE 1st
OBSERVATION...

$\ldots X_2$ IS THE 2nd
OBSERVATION...

$\ldots$ AND X_n IS OUR
FINAL OBSERVATION
IN A LIST THAT HAS n
OBSERVATIONS IN IT.

SAMPLE SIZE (n)

FROM P. 54

THE SAMPLE SIZE IS THE TOTAL NUMBER OF MEASUREMENTS INCLUDED IN A SINGLE
SAMPLE. IN GENERAL, A LARGER n INCREASES THE **CONFIDENCE** WE CAN HAVE IN
OUR STATISTICAL CONCLUSIONS, BUT **ONLY** IF OUR SAMPLE IS **RANDOM**!

SAMPLE AVERAGE ($\bar{x}$)

FROM P. 59

WE COMPUTE THE **AVERAGE** IN A SAMPLE BY ADDING UP ALL VALUES IN THAT SAMPLE
AND DIVIDING BY THE SAMPLE SIZE. HERE'S THE FORMULA:

ARRRRR.
WE CALL
OUR SAMPLE
AVERAGE
"XBAR."

$$\bar{X} = \frac{X_1 + X_2 + \ldots + X_n}{n}$$

THE AVERAGE IS ALSO COMMONLY KNOWN AS THE "**ARITHMETIC MEAN**," OR
JUST "**THE MEAN**" FOR SHORT. IN THIS BOOK WE'VE AVOIDED "MEAN" AND USED
"AVERAGE" INSTEAD BECAUSE WE HOPE THAT BY USING THIS MORE FAMILIAR TERM
WE CAN HELP MAKE STATISTICAL INFERENCE FEEL MORE FAMILIAR. ALSO, WE BELIEVE
MOST READERS THINK OF THE ARITHMETIC MEAN WHEN THEY HEAR THE WORD
"AVERAGE" ANYWAY.

WHATEVER YOU CALL IT, THE AVERAGE IS THE MOST BASIC MEASURE OF THE
CENTRAL TENDENCY IN A DISTRIBUTION. THERE ARE SEVERAL OTHER WAYS TO
REFINE OUR UNDERSTANDING OF HOW A PARTICULAR DATA SET CLUMPS TOGETHER,
BUT THE CHOICE OF WHICH TO USE DEPENDS ON THE SITUATION.

FOR EXAMPLE, THE **MEDIAN** IS THE "MIDDLE VALUE" OF A SAMPLE AND MAY BE
PREFERABLE IN CASES OF SKEW. SIMILARLY, A **TRIMMED AVERAGE** IS COMPUTED
BY EXCLUDING A SMALL PERCENTAGE OF THE SMALLEST AND LARGEST VALUES, AND
MAY BE PREFERABLE WHEN THERE ARE EXTREME VALUES IN A SAMPLE.

FROM P. 65

STANDARD DEVIATION (s)

OUR GOAL WHEN WE CALCULATE **STANDARD DEVIATION** IS
TO GET A SENSE OF THE AVERAGE DISTANCE FROM THE AVERAGE VALUE.
HERE'S HOW TO DO IT IN (MOSTLY) PLAIN ENGLISH:

1) CALCULATE THE DISTANCE BETWEEN EACH MEASUREMENT x AND THE SAMPLE
AVERAGE $\bar{x}$. WE CALL THIS DISTANCE A **DEVIATION**.

2) SQUARE EACH DEVIATION.

3) ADD UP ALL THE SQUARED DEVIATIONS.

4) DIVIDE THE SUM BY $n - 1$ (IF WE STOP HERE,
WE GET WHAT'S CALLED THE **VARIANCE**.)

5) TAKE THE SQUARE ROOT OF THE WHOLE SHEBANG.

HERE'S THE ACTUAL FORMULA:

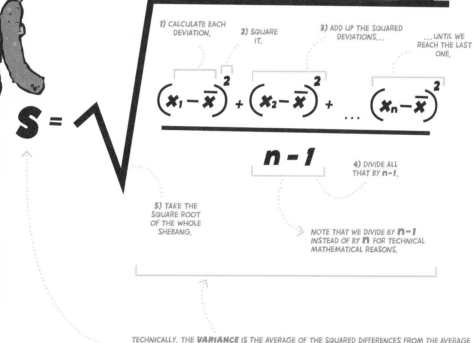

1) CALCULATE EACH
DEVIATION.

2) SQUARE
IT.

3) ADD UP THE SQUARED
DEVIATIONS...

...UNTIL WE
REACH THE LAST
ONE.

$$s = \sqrt{\frac{\left(x_1 - \bar{x}\right)^2 + \left(x_2 - \bar{x}\right)^2 + \ldots \left(x_n - \bar{x}\right)^2}{n-1}}$$

4) DIVIDE ALL
THAT BY $n-1$.

5) TAKE THE
SQUARE ROOT
OF THE WHOLE
SHEBANG.

NOTE THAT WE DIVIDE BY $n-1$
INSTEAD OF BY n FOR TECHNICAL
MATHEMATICAL REASONS.

TECHNICALLY, THE **VARIANCE** IS THE AVERAGE OF THE SQUARED DIFFERENCES FROM THE AVERAGE,
AND THE **STANDARD DEVIATION** IS THE SQUARE ROOT OF THE VARIANCE. NOTE THAT WE USE THE
SINGLE LETTER **S** TO REFER SPECIFICALLY TO THE STANDARD DEVIATION OF OUR SAMPLE.

FROM P. 70

VARIABLE (X)

A VARIABLE IS A PARTICULAR QUALITY WE'RE CURIOUS ABOUT.
HOWEVER, BECAUSE IN STATISTICS WE ALWAYS COLLECT DATA
RANDOMLY, WE REFER TO THE VARIABLES WE'RE LOOKING AT AS
RANDOM VARIABLES. TECHNICALLY, A **RANDOM VARIABLE**
IS A VARIABLE WHOSE VALUE IS RANDOM.

IN THE SHORT TERM, WE HAVE NO WAY OF PREDICTING A
RANDOM VARIABLE'S VALUE BEFORE WE GATHER IT. IT'S LIKE
A COIN FLIP. IN THE LONG TERM, WE PREDICT THE VALUE OF A
RANDOM VARIABLE USING **PROBABILITY** (SEE BELOW).

WORM LENGTH IS A
VARIABLE.

SO IS **PIRATE
INCOME**.

SO IS **DRAGON
SPEED**.

215

DISTRIBUTIONS

IN GENERAL MATHEMATICAL TERMS, THE WORD **DISTRIBUTION** DESCRIBES THE ARRANGEMENT OF ALL THE POSSIBLE VALUES FOR A RANDOM VARIABLE. IF, FOR EXAMPLE, YOU MADE A HISTOGRAM OF ALL THE VALUES OF A VARIABLE IN AN ENTIRE POPULATION, YOU'D BE LOOKING AT THE **POPULATION DISTRIBUTION** FOR THAT VARIABLE.

MORE GENERALLY, DISTRIBUTIONS ALLOW US TO COMPUTE **PROBABILITIES** (OR LONG-RUN LIKELIHOODS) OF RANDOMLY GRABBING VALUES FROM PARTICULAR INTERVALS. IN STATISTICAL INFERENCE, WE CALCULATE PROBABILITIES USING **SAMPLING DISTRIBUTIONS** (SEE BELOW), BUT IF WE HAD A POPULATION DISTRIBUTION IN FRONT OF US, WE COULD ALSO USE IT TO CALCULATE PROBABILITIES. HERE'S HOW:

IF WE SOMEHOW KNEW HOW THE **ENTIRE POPULATION** OF FISH IN A LAKE, SORTED BY LENGTH, WAS DISTRIBUTED...

...WE COULD DO SOME MATH TO CALCULATE THE PROPORTION OF FISH INSIDE ANY AREA OF THAT DISTRIBUTION...

...LIKE **THIS AREA** COVERING THE RANGE FROM 8 TO 12 INCHES.

IT FOLLOWS THAT IF WE REACHED INTO THE LAKE AND **RANDOMLY GRABBED ONE FISH**...

...THE **PROBABILITY** THAT IT WOULD HAVE A LENGTH BETWEEN 8 AND 12 INCHES IS THE SAME AS THE PROPORTION OF THE TOTAL DISTRIBUTION THAT'S INSIDE THE DARKER AREA.

IF **HALF** OF ALL THE FISH ARE BETWEEN 8 AND 12 INCHES...

...THE PROBABILITY I'LL RANDOMLY CATCH ONE IN THAT RANGE IS 50%, OR **0.5**.

0 1 2 3 4 5 6 7 8 9 10 11 12 13 14 15 16

OF COURSE, IN REALITY, WE NEVER ACTUALLY GET TO LOOK AT AN ENTIRE POPULATION DISTRIBUTION. IF WE DID, WE WOULDN'T NEED STATISTICS.

SAMPLE STATISTICS VS. POPULATION PARAMETERS

SINCE OUR GOAL IN STATISTICS IS ALWAYS TO USE **SAMPLES** TO MAKE GUESSES ABOUT **POPULATIONS**, WE HAVE DIFFERENT TERMS AND TECHNICAL NOTATION FOR EACH.

WE CALL QUALITIES IN A SAMPLE **"STATISTICS."**

WE CALL QUALITIES IN A POPULATION **"PARAMETERS."**

WHEN WE'RE WRITING FORMULAS, **XBAR** REFERS EXCLUSIVELY TO OUR **SAMPLE AVERAGE:**

THE LOWERCASE GREEK LETTER **MU** REFERS EXCLUSIVELY TO THE **POPULATION AVERAGE:**

S REFERS EXCLUSIVELY TO OUR **SAMPLE STANDARD DEVIATION:**

THE LOWERCASE GREEK LETTER **SIGMA** REFERS EXCLUSIVELY TO THE **POPULATION STANDARD DEVIATION:**

STATISTICS ARE THE THINGS WE ACTUALLY MEASURE AND THEREFORE KNOW WITH CERTAINTY.

PARAMETERS ARE THE THINGS WE REALLY WANT TO KNOW, BUT CAN ONLY MAKE GUESSES ABOUT.

THE NORMAL DISTRIBUTION

FROM P. 94

IN MATHEMATICS AND PROBABILITY THEORY, THERE ARE LOTS OF DIFFERENT KINDS OF DISTRIBUTIONS THAT COME IN LOTS OF DIFFERENT SHAPES. BY FAR THE MOST FAMOUS HOWEVER, IS THE **NORMAL DISTRIBUTION.** IN STATISTICS, WE CARE MOST ABOUT IT BECAUSE IT'S HOW AVERAGES TEND TO PILE UP (SEE THE **CLT**, BELOW).

LIKE ANY OTHER DISTRIBUTION, WE CAN CARVE UP A NORMAL DISTRIBUTION INTO AREAS THAT DEPICT PROBABILITIES FOR THE VALUES INSIDE IT. WE LEARN HOW TO DO THIS ON PAGE 115, BUT HERE'S AN EXAMPLE:

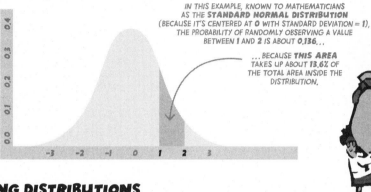

IN THIS EXAMPLE, KNOWN TO MATHEMATICIANS AS THE **STANDARD NORMAL DISTRIBUTION** (BECAUSE IT'S CENTERED AT 0 WITH STANDARD DEVIATION = 1), THE PROBABILITY OF RANDOMLY OBSERVING A VALUE BETWEEN 1 AND 2 IS ABOUT 0.136...

...BECAUSE **THIS AREA** TAKES UP ABOUT 13.6% OF THE TOTAL AREA INSIDE THE DISTRIBUTION.

SAMPLING DISTRIBUTIONS

FROM P. 95

TECHNICALLY, A **SAMPLING DISTRIBUTION** IS THE DISTRIBUTION OF A SAMPLE STATISTIC. ALTHOUGH WE CAN BUILD SAMPLING DISTRIBUTIONS FOR ANY STATISTIC (STANDARD DEVIATIONS, MEDIANS, ETC.) WE'RE FOCUSING HERE ON SAMPLING DISTRIBUTIONS MADE OF AVERAGES. SO, FOR EXAMPLE, IF WE COLLECTED MANY, MANY SAMPLES OF SIZE n FROM A POPULATION, COMPUTED $\bar{x}$ FOR EACH, THEN MADE A HISTOGRAM OF ALL THE $\bar{x}$ VALUES, WE'D BE LOOKING AT THE SAMPLING DISTRIBUTION OF $\bar{x}$. THE PILE IN CRAZY BILLY'S BAIT BARN IS AN EXAMPLE (SEE PAGE 107). SAMPLING DISTRIBUTIONS ARE **KEY** TO STATISTICAL INFERENCE.

THE CENTRAL LIMIT THEOREM (CLT)

FROM P. 101–102

MUCH OF STATISTICAL INFERENCE DEPENDS ON THE **CENTRAL LIMIT THEOREM**, WHICH STATES THAT THE SAMPLING DISTRIBUTION OF $\bar{x}$ BECOMES APPROXIMATELY **NORMAL** AS THE SAMPLE SIZE n GETS LARGE.

MORE SPECIFICALLY, FOR **RANDOM SAMPLES** OF LARGE SIZE n TAKEN FROM A SINGLE POPULATION WITH AVERAGE μ AND SD σ, THE DISTRIBUTION OF $\bar{x}$ IS APPROXIMATELY **NORMAL** WITH AVERAGE μ AND SD EQUAL TO $\sigma/\sqrt{n}$.

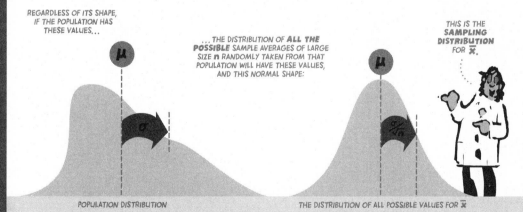

REGARDLESS OF ITS SHAPE, IF THE POPULATION HAS THESE VALUES...

...THE DISTRIBUTION OF **ALL THE POSSIBLE** SAMPLE AVERAGES OF LARGE SIZE n RANDOMLY TAKEN FROM THAT POPULATION WILL HAVE THESE VALUES, AND THIS NORMAL SHAPE:

THIS IS THE **SAMPLING DISTRIBUTION** FOR $\bar{x}$.

POPULATION DISTRIBUTION

THE DISTRIBUTION OF ALL POSSIBLE VALUES FOR $\bar{x}$

$\sigma/\sqrt{n}$ IS ALSO KNOWN AS THE **STANDARD ERROR.**

217

THE CENTRAL LIMIT THEOREM (CONT.)

THE CLT IS A VERY GENERAL RESULT THAT WILL ALMOST ALWAYS APPLY AS DESCRIBED IN THE BOOK. THAT SAID, THERE ARE **IMPORTANT CONDITIONS** UNDERLYING THE CLT.

FIRST, THE CLT ONLY WORKS IF EACH OF THE VALUES FOR $x_1, x_2, x_3 \ldots x_n$ IN OUR SAMPLE COMES FROM THE **SAME EXACT POPULATION DISTRIBUTION**. THIS WILL USUALLY BE TRUE FOR SAMPLES OBTAINED IN PRACTICE, BUT CAN BE RELEVANT IF WE'RE INVESTIGATING MORE COMPLICATED QUESTIONS.

SECOND, EACH MEASUREMENT x_i HAS TO BE **RANDOM**. TECHNICALLY THIS ALSO MEANS THAT ALL VALUES FOR x_i HAVE TO BE **INDEPENDENT** OF ONE ANOTHER, SO THAT THE VALUE OF EACH MEASUREMENT x_i DOES NOT DEPEND ON THE VALUES OF THE OTHER SAMPLE VALUES. FOR EXAMPLE, MEASUREMENTS OF TEMPERATURE TAKEN ACROSS A GEOGRAPHICAL REGION WILL **NOT** BE INDEPENDENT, SINCE THE TEMPERATURE AT ONE LOCATION WILL TEND TO BE SIMILAR TO THE TEMPERATURE AT A NEARBY LOCATION; STATISTICIANS WOULD SAY THESE MEASUREMENTS ARE "**CORRELATED**," BECAUSE THERE EXISTS A SYSTEMATIC UNDERLYING PATTERN THAT INFLUENCES THE VALUE OF EACH x_i (SEE **CORRELATION**, BELOW.)

FINALLY, AND MOST TECHNICALLY, THE CENTRAL LIMIT THEOREM APPLIES WHEN n APPROACHES INFINITY, BUT FOR PRACTICAL PURPOSES WE USE AN **APPROXIMATE VERSION** OF THE CLT THAT WORKS WHEN $n \geq 30$. AS A RESULT, IN PRACTICE WE CONSIDER ANY SAMPLE SIZE $n \geq 30$ TO BE "**LARGE**." THIS OFTEN FEELS ARBITRARY, BUT A MORE THOROUGH EXPLANATION WOULD REQUIRE LOTS MORE MATH.

FROM P. 112

PROBABILITIES

IN THE BOOK WE NOTE PROBABILITIES AS PERCENTAGES (E.G., 95%), BUT IN MATHEMATICS WE USE NUMBERS BETWEEN 0 AND 1 TO EXPRESS THE SAME THING (E.G., 95% = 0.95). SO, FORMALLY, A **PROBABILITY** IS A NUMBER BETWEEN 0 AND 1 THAT QUANTIFIES THE LIKELIHOOD THAT A RANDOM EVENT WILL OCCUR; THE CLOSER THE PROBABILITY IS TO 1 (OR 100%), THE MORE LIKELY THE EVENT IS TO OCCUR, IN THE LONG RUN. IN OTHER WORDS, PROBABILITIES ARE LIKE PREDICTIONS ABOUT THE LONG RUN. THE TRICKY THING ABOUT THEM, HOWEVER, IS THAT THEY **ONLY REFER TO THE LONG RUN**.

IF, FOR EXAMPLE, THERE ARE EQUAL NUMBERS OF MALE AND FEMALE VOTERS IN A STATE, THE PROBABILITY THAT A RANDOMLY SELECTED VOTER IS FEMALE IS **0.5**. HOWEVER, THE FIRST FEW VOTERS RANDOMLY SAMPLED MAY WELL BE ALL MALE, JUST BY CHANCE. THE 0.5 SPEAKS TO WHAT WOULD HAPPEN **IN THE LONG RUN**: IF WE RANDOMLY SAMPLE ENOUGH VOTERS, WE WILL **EVENTUALLY** END UP WITH **ROUGHLY EQUAL NUMBERS** OF MALE AND FEMALE VOTERS.

IN ANOTHER EXAMPLE, WHEN WE FLIP A COIN, THERE'S A PROBABILITY OF 0.5 THAT IT WILL LAND ON HEADS. BUT EVEN IF WE JUST FLIPPED THE COIN ONCE AND GOT HEADS, THE PROBABILITY OF THE NEXT FLIP LANDING ON HEADS IS **STILL 0.5**. IN THIS WAY, EACH FLIP IS **INDEPENDENT** OF THE OTHERS.

IN SUM, ANY TIME WE CALCULATE A PROBABILITY, IT CAN BE EXPRESSED AS A NUMBER BETWEEN 0 AND 1 (OR, EQUIVALENTLY, 0 AND 100%), AND THAT NUMBER ALWAYS CORRESPONDS TO THE AREA INSIDE A PROBABILITY DISTRIBUTION. BY DEFINITION, THE TOTAL AREA INSIDE ANY PROBABILITY DISTRIBUTION EQUALS **1**.

FROM P. 114

PROBABILITY MATH

TECHNICALLY, WE CAN COMPUTE AREAS INSIDE ANY DISTRIBUTION (LIKE THE NORMAL ONE DEPICTED ON PAGE 114) USING **INTEGRATION**, WHICH IS A **CALCULUS TECHNIQUE**. IN PRACTICE, STATISTICIANS ASK COMPUTERS TO DO THE CALCULATIONS FOR THEM.

IN THE BILLY'S BAIT BARN EXAMPLE, THE SAMPLING DISTRIBUTION IS NORMAL-SHAPED BECAUSE OF THE CENTRAL LIMIT THEOREM. HOWEVER, IN A LOT OF OTHER STATISTICAL APPLICATIONS, A PARTICULAR SAMPLING DISTRIBUTION WON'T BE NORMAL-SHAPED, BUT WE CAN STILL DO CALCULATIONS LIKE THESE, USING CALCULUS. FOR MORE ABOUT THAT, SEE CHAPTER 14.

ALL DISTRIBUTIONS CAN BE DRAWN AS CURVES, BUT THEY CAN ALSO BE WRITTEN AS **FUNCTIONS**, WHICH ARE LIKE MATH MACHINES THAT TAKE **INPUTS** (IN THIS CASE A RANDOM VARIABLE) AND TURN THEM INTO **OUTPUTS** (IN THIS CASE A PROBABILITY).

IN **MATH NOTATION**, HERE'S A GENERIC WAY TO WRITE ABOUT A PROBABILITY FUNCTION f WITH AVERAGE μ AND STANDARD DEVIATION σ :

IF X IS A **DISCRETE RANDOM VARIABLE** WITH DISTRIBUTION $f_{\mu,\sigma}$...

... THEN $f_{\mu,\sigma}(x)$ EQUALS THE PROBABILITY THAT X TAKES THE VALUE x.

UNFORTUNATELY, IT GETS EVEN MORE COMPLICATED FAST. FOR EXAMPLE, HERE'S THE PROBABILITY FUNCTION FOR THE NORMAL DISTRIBUTION:

$$h_{\mu,\sigma}(x) = \frac{1}{\sigma\sqrt{2\pi}}\exp\left\{-\frac{1}{2\sigma^2}(x-\mu)^2\right\}$$

THOUGH THIS NOTATION IS **TERRIFYING** AT FIRST GLANCE, IN THE SCOPE OF BROADER STATISTICAL AND MATHEMATICAL INQUIRY, PROBABILITY FUNCTIONS LIKE THIS ARE ENORMOUSLY USEFUL BECAUSE THEY RELATE **PARTICULAR KINDS OF RANDOM EVENTS** (LIKE CATCHING A CERTAIN SIZE FISH) WITH **PREDICTABLE LONG-RUN OUTCOMES** (HOW OFTEN YOU'D EXPECT THAT TO HAPPEN IN THE LONG RUN).

FROM P. 129

ESTIMATING A SAMPLING DISTRIBUTION

IN PRACTICE, WHEN WE MAKE USE OF THE CLT, WE HAVE NO WAY OF KNOWING THE REAL VALUES FOR THE PARAMETERS μ AND σ, SO WE USE THE STATISTICS $\bar{X}$ AND S TO **APPROXIMATE** THEM. THIS APPROXIMATION WORKS BECAUSE WE GATHER OUR STATISTICS RANDOMLY. AS A RESULT, WE EXPECT $\bar{X}$ TO DIFFER FROM μ AND S TO DIFFER FROM σ, BUT **ONLY BECAUSE OF CHANCE VARIATION**.

AFTER WE'VE SWAPPED IN THE APPROXIMATE VALUES, WE CALL THE RESULT AN **ESTIMATED SAMPLING DISTRIBUTION**:

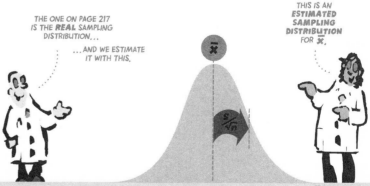

THE ONE ON PAGE 217 IS THE **REAL** SAMPLING DISTRIBUTION...

...AND WE ESTIMATE IT WITH THIS.

THIS IS AN **ESTIMATED SAMPLING DISTRIBUTION** FOR $\bar{X}$.

$\bar{x}$

$\frac{s}{\sqrt{n}}$

AN **ESTIMATED** DISTRIBUTION OF ALL POSSIBLE VALUES FOR $\bar{x}$

NOTE THAT WE CAN ALSO BUILD **ESTIMATED SAMPLING DISTRIBUTIONS** FOR OTHER STATISTICS, SUCH AS S (SEE PAGE **201**, FIRECRACKERS), BUT WE CAN ONLY EXPECT A SAMPLING DISTRIBUTION TO BE NORMAL-SHAPED WHEN THE CLT OR SIMILAR RESULTS APPLY.

FROM P. 135

CONFIDENCE INTERVALS

TECHNICALLY, A **CONFIDENCE INTERVAL** IS A TYPE OF INTERVAL ESTIMATE THAT
RELATES TO A PARTICULAR **CONFIDENCE LEVEL**. CONFIDENCE INTERVALS CAN
BE COMPUTED FOR ANY PARAMETER, ALTHOUGH THE SPECIFIC TECHNICAL DETAILS
WILL CHANGE. HERE'S THE FORMULA FOR HOW TO COMPUTE A **95% CONFIDENCE
INTERVAL** FOR A POPULATION AVERAGE μ :

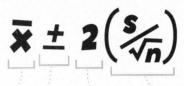

$$\bar{x} \pm 2\left(\frac{s}{\sqrt{n}}\right)$$

WHEN STATISTICIANS TALK
ABOUT THIS WHOLE FORMULA THEY SAY
"ESTIMATE PLUS OR MINUS CUTOFF,
TIMES SD OF ESTIMATE."

WE USE $\bar{x}$ TO ESTIMATE
THE VALUE OF THE
POPULATION AVERAGE.

THE PLUS OR MINUS
MEANS WE GO OUT FROM
THE MIDDLE, IN BOTH
DIRECTIONS.

WE CALL THIS THE
CUTOFF. IT TELLS US HOW
FAR OUT IN THE TAILS OF
THE DISTRIBUTION TO GO
TO CAPTURE WHATEVER SIZE
PROBABILITY WE WANT.

THIS IS HOW
WE ESTIMATE
THE SD OF $\bar{x}$.

HERE'S THE CONCLUSION
WE CAN DRAW FROM
THAT FORMULA.

WE'RE 95%
CONFIDENT...

...THAT μ IS
SOMEWHERE
INSIDE THIS
RANGE.

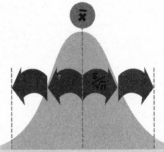

AN **ESTIMATED** SAMPLING DISTRIBUTION FOR $\bar{x}$

WE CAN **CHANGE OUR CONFIDENCE LEVEL** BY **CHANGING THE CUTOFF**. FOR EXAMPLE,
IF WE WANTED AN 80% CONFIDENCE INTERVAL FOR THE POPULATION AVERAGE, WE WOULD USE **1.3**
AS OUR CUTOFF, SINCE APPROXIMATELY **80%** OF A NORMAL DISTRIBUTION IS CONTAINED WITHIN
1.3 STANDARD DEVIATIONS OF THE CENTER. (FOR AN EXAMPLE, SEE PAGE **157.**)

IDEALLY, WE WANT THE **NARROWEST POSSIBLE INTERVAL** FOR ANY LEVEL OF
CONFIDENCE, SINCE A NARROWER INTERVAL IS MORE PRECISE. ONE SUREFIRE WAY TO GET
A NARROWER INTERVAL IS TO INCREASE **n** (BY COLLECTING MORE OBSERVATIONS). THAT'S
WHY A BIGGER SAMPLE SIZE IS BETTER! (FOR AN EXAMPLE, SEE PAGE **159.**)

REMEMBER THAT OUR **LEVEL OF CONFIDENCE** IS BASED ON A PROBABILITY VALUE,
SO IT'S ONLY RELEVANT WHEN WE THINK ABOUT THE LONG RUN. AS A RESULT, WHEN WE
COMPUTE AN INTERVAL USING THE FORMULA ABOVE, **WE DON'T KNOW WHETHER IT
ACTUALLY CONTAINS μ OR NOT!** ALL WE CAN SAY IS THAT INTERVALS CONSTRUCTED
IN THIS WAY WILL TEND TO BE ACCURATE IN THE LONG RUN. FOR A **95%** CONFIDENCE
INTERVAL WE CAN EXPECT TO BE WRONG 5% OF THE TIME... IN THE LONG RUN.

FROM P. 163

HYPOTHESIS TESTS

HYPOTHESIS TESTING USES THE **SAME UNDERLYING STATISTICAL MACHINERY** THAT WE USE WHEN WE COMPUTE A CONFIDENCE INTERVAL. WE STILL START BY BUILDING AN ESTIMATED SAMPLING DISTRIBUTION. THIS TIME, HOWEVER, WE USE IT TO QUESTION WHETHER WE THINK A **PARTICULAR VALUE** FOR THE POPULATION PARAMETER IS TRUE OR NOT. WE DO THIS BY ASKING **HOW CONSISTENT** OUR OBSERVED DATA ARE WITH THAT PARTICULAR VALUE.

FORMALLY, HYPOTHESIS TESTS START WITH TWO HYPOTHESES. ONE IS OUR **RESEARCH HYPOTHESIS** (SOMETIMES CALLED THE **ALTERNATE HYPOTHESIS**) AND THE OTHER IS THE **NULL HYPOTHESIS** (IN THE BOOK WE USE THE WORD "DULL").

HYPOTHESIS TESTS ALWAYS END WHEN WE CALCULATE A **P-VALUE** AND USE IT TO MAKE A **FORMAL DECISION** ABOUT WHETHER WE THINK OUR STATISTIC IS FAR ENOUGH AWAY FROM THE PARAMETER PREDICTED BY THE NULL HYPOTHESIS TO JUSTIFY REJECTING THE NULL HYPOTHESIS IN FAVOR OF ANOTHER EXPLANATION.

HERE IS A QUICK SUMMARY OF THE UNDERLYING LOGIC:

OUR NULL HYPOTHESIS BOILS DOWN TO THIS.

IF μ IS, IN REALITY, LOCATED **RIGHT HERE**...

μ

... WE'RE **VERY UNLIKELY** IN THE LONG RUN...

... TO RANDOMLY GRAB VALUES FOR $\bar{x}$ WAY OUT IN THE ENDS.

SO IF THE $\bar{x}$ WE ACTUALLY FOUND IS WAY OUT IN THE ENDS, WITH A P-VALUE OF LESS THAN 0.05, MAYBE THE NULL HYPOTHESIS IS FALSE.

HMMMMM.

AN **ESTIMATED** SAMPLING DISTRIBUTION FOR $\bar{x}$, ASSUMING THE NULL HYPOTHESIS IS TRUE

IN THIS BOOK WE'VE FOCUSED ON HYPOTHESIS TESTS ABOUT AVERAGES. IN PRACTICE, THESE SAME GENERAL STEPS CAN WORK FOR **ANY PARAMETER** AND ITS CORRESPONDING STATISTIC, BUT THE MATH DETAILS WILL VARY.

FROM P. 169

P-VALUES

FORMALLY, A **P-VALUE** CAN BE DEFINED AS THE PROBABILITY THAT WE WOULD OBSERVE DATA AT LEAST AS EXTREME AS THE DATA WE ACTUALLY OBSERVED IF THE NULL HYPOTHESIS WERE TRUE. WHEW! IN THE BOOK WE INDICATE P-VALUES WITH PERCENTAGES, BUT AGAIN IT'S COMMON TO USE NUMBERS BETWEEN 0 AND 1. A P-VALUE OF 5% IS COMMONLY EXPRESSED BY THE NUMBER 0.05.

SOMETIMES WE CALCULATE A P-VALUE FOR **BOTH ENDS** OF OUR ESTIMATED SAMPLING DISTRIBUTION (SEE PAGE 181; THIS IS CALLED A **TWO-TAILED TEST**), AND SOMETIMES WE CALCULATE A P-VALUE FOR **ONE END** ONLY (SEE PAGE 187; THIS IS CALLED A **ONE-TAILED TEST**). THE CHOICE OF WHICH TO USE DEPENDS ON WHAT SORT OF RESEARCH HYPOTHESIS WE'RE CURIOUS ABOUT.

IN PRACTICE, STATISTICIANS USE COMPUTERS TO **CALCULATE A P-VALUE** (CALCULUS AGAIN). HOWEVER, IT CAN BE HELPFUL TO NOTE THAT (WHEN WE'RE PERFORMING A TWO-TAILED TEST) A PROBABILITY OF 0.05 IS PRECISELY THAT AREA THAT **DOES NOT FIT** INSIDE A 95% CONFIDENCE INTERVAL. AS A RESULT, ONE RELATIVELY SIMPLE WAY TO CARRY OUT A HYPOTHESIS TEST IS TO CONSTRUCT A 95% CONFIDENCE INTERVAL FOR μ AS DESCRIBED ABOVE. IF THE VALUE OF μ PREDICTED BY THE NULL HYPOTHESIS ISN'T INSIDE THAT INTERVAL, THE P-VALUE MUST BE LESS THAN 0.05.

ON A RELATED NOTE: INCREASES IN n RESULT IN SMALLER P-VALUES. THAT'S WHY, IN STATS JARGON, COLLECTING MORE OBSERVATIONS IS A SUREFIRE WAY TO GET MORE **POWER** TO REJECT A NULL HYPOTHESIS. IT'S ANOTHER REASON WHY A BIGGER SAMPLE SIZE IS BETTER!

P-VALUES (CONT.)

REMEMBER THAT A P-VALUE IS A MEASURE OF PROBABILITY, SO IT'S ONLY RELEVANT WHEN WE THINK ABOUT THE LONG RUN.

IN PRACTICE, WE REJECT THE NULL HYPOTHESIS IF OUR P-VALUE IS "SUFFICIENTLY SMALL," WHICH (USING A COMMON RULE OF THUMB) MEANS **LESS THAN 0.05**, BUT THERE'S NOTHING MAGICAL ABOUT THAT NUMBER. A PROBABILITY OF "LESS THAN 0.05" MEANS THE SAME THING AS "FEWER THAN 1 OUT OF EVERY 20 TIMES IN THE LONG RUN."

SO FOR EXAMPLE, IF WE PERFORM A HYPOTHESIS TEST AND GET A P-VALUE OF **0.049**, IT MEANS THAT "IF THE NULL HYPOTHESIS WERE TRUE, WE'D EXPECT, JUST BY CHANCE, TO SEE DATA LIKE OURS ABOUT **49** OUT OF EVERY **1,000** TIMES IN THE LONG RUN." BECAUSE 49/1,000 IS LESS THAN 1/20, WE'D CONCLUDE THAT OUR DATA DON'T MATCH THAT NULL HYPOTHESIS WELL.

FROM P. 172 ## WE ALWAYS MIGHT BE WRONG

ALL STATISTICAL INQUIRY IS BASED ON RANDOM SAMPLING, AND **ALL** STATISTICAL INFERENCE IS BASED ON CALCULATING PROBABILITIES. AS A RESULT, ANY TIME WE USE A SAMPLE STATISTIC TO MAKE A GUESS ABOUT A POPULATION PARAMETER WE **MIGHT BE WRONG!**

BECAUSE OF THIS FACT, WE HAVE TO BE **VERY CAREFUL** ABOUT THE LANGUAGE WE USE WHEN WE'RE TEMPTED TO MAKE **TRUTH CLAIMS** BASED ON STATISTICS. WE HAVE TO BE ESPECIALLY CAREFUL WHEN WE'RE MAKING FORMAL CONCLUSIONS BASED ON P-VALUES, BECAUSE WE ONLY USE P-VALUES WHEN WE'RE INVESTIGATING THEORIES THAT WE'RE EXCITED ABOUT.

IF WE'RE INVESTIGATING A THEORY AND WE USE A SMALL P-VALUE TO ADD SUPPORT TO IT, WE **MIGHT BE WRONG**. OUR THEORY MIGHT BE WRONG AND CHANCE VARIATION MIGHT BE A BETTER EXPLANATION FOR OUR RESULTS. STATISTICIANS CALL THIS A **FALSE POSITIVE**, OR **TYPE 1** ERROR.

ALTERNATIVELY, IF WE'RE INVESTIGATING A THEORY AND WE USE A LARGER P-VALUE TO REJECT IT, WE **MIGHT BE WRONG**. OUR THEORY MIGHT ACTUALLY BE TRUE AND WE GOT RESULTS CLOSE TO THE VALUE PREDICTED BY THE NULL JUST BY CHANCE. STATISTICIANS CALL THIS A **FALSE NEGATIVE**, OR **TYPE 2** ERROR.

IN SUM, HYPOTHESIS TESTS ARE ONLY ABOUT ASKING THE QUESTION, "HOW LIKELY IS IT THAT WE JUST GOT OUR RESULTS BY CHANCE?" THEY CAN'T BE USED TO DISPROVE OR PROVE ANY THEORY CONCLUSIVELY; THEY CAN **ONLY** BE USED TO HELP US **CHALLENGE A NULL HYPOTHESIS**.

IN STATISTICS WE **ALWAYS MIGHT BE WRONG**. THIS IS ALWAYS THE CASE. IT'S A RESULT OF THE FACT THAT WE'RE USING A LONG-RUN PORTRAIT TO EVALUATE SHORT-TERM OBSERVATIONS.

FROM P. 197

INFERENCE ON A DIFFERENCE

TO CALCULATE A CONFIDENCE INTERVAL ABOUT THE **DIFFERENCE** BETWEEN TWO POPULATION AVERAGES, WE CAN USE A FORMULA THAT'S ONLY A BIT DIFFERENT FROM THE ONE WE LEARNED ABOVE.

IN THIS CASE, WE'RE CURIOUS ABOUT THE **DIFFERENCE** BETWEEN TWO POPULATION AVERAGES, AND WE ESTIMATE THAT DIFFERENCE WITH TWO SAMPLE AVERAGES.

THIS IS HOW WE **COMBINE THE VARIABILITY** OF THE TWO POPULATIONS.

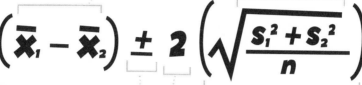

$$\left(\bar{x}_1 - \bar{x}_2\right) \pm 2\left(\sqrt{\frac{S_1^2 + S_2^2}{n}}\right)$$

THE PLUS OR MINUS MEANS WE GO OUT FROM THE MIDDLE, IN BOTH DIRECTIONS.

WE USE A **CUTOFF** OF 2 IF WE WANT A 95% CONFIDENCE INTERVAL, BUT OF COURSE THIS CAN BE CHANGED.

THIS WHOLE THING IS THE **SD OF OUR ESTIMATED SAMPLING DISTRIBUTION.** IN THIS STORY IT EQUALS APPROXIMATELY 1.

IN OUR STORY, $\bar{x}_1 = 59.7$, $S_1 = 4.6$, $\bar{x}_2 = 44.2$, AND $S_2 = 4.7$.

NOTE THAT THIS FORMULA CAN REQUIRE OTHER TWEAKS AS WELL. WE HAVE TO TWEAK IT, FOR EXAMPLE, IF WE HAVE **DIFFERENT SAMPLE SIZES FOR OUR TWO SAMPLES**, OR IF THEY'RE **TOO SMALL TO YIELD A NORMAL-SHAPED SAMPLING DISTRIBUTION.**

FROM P. 199

INFERENCE WITH A SMALL SAMPLE SIZE

WHENEVER WE'RE DOING INFERENCE ON A POPULATION AVERAGE AND WE HAVE A **SMALL SAMPLE SIZE** (E.G., WHEN $n < 30$), WE CAN'T RELY ON THE **CLT**, SO WE USE WHAT'S CALLED THE **T-DISTRIBUTION**, WHICH WORKS **ONLY** IF THE POPULATION ITSELF IS NORMAL.

FOR MURKY HISTORICAL REASONS RELATED TO ITS CODISCOVERY BY A GUY AT THE GUINNESS BREWING COMPANY, THE **T-DISTRIBUTION** IS ALSO KNOWN AS THE "**STUDENTS DISTRIBUTION.**"

THE T-DISTRIBUTION IS FATTER THAN THE STANDARD NORMAL DISTRIBUTION, AND IN PRACTICE, WE MODIFY THE T-DISTRIBUTION DEPENDING ON THE SIZE OF n (A SMALLER n REQUIRES A FATTER **T**). AS A RESULT, WHEN WE USE THE T-DISTRIBUTION INSTEAD OF THE NORMAL DISTRIBUTION, OUR CONFIDENCE INTERVALS WILL BE WIDER, AND OUR P-VALUES WILL BE BIGGER. IN BOTH CASES, WE REQUIRE GREATER STATISTICAL EVIDENCE TO ACHIEVE THE SAME LEVEL OF CONFIDENCE. IT'S LIKE WE'RE BEING PENALIZED FOR THE SMALL SAMPLE SIZE.

IN OUR STORY, THE SPECIFIC T-DISTRIBUTION WE USED IS T_9, WHICH STATISTICIANS CALL THE "T WITH 9 [AS IN $n-1$] DEGREES OF FREEDOM."

HERE ARE THE PH MEASUREMENTS FROM OUR 10 RANDOM ALIENS: **2.09, 2.39, 1.32, 2.99, 2.62, 2.60, 2.45, 2.13, 2.27, 2.95.** FROM THESE NUMBERS WE CAN CALCULATE $\bar{x} = 2.38$, $S = 0.48$, AND $n = 10$, WHICH WE PLUG INTO THE FOLLOWING FORMULA TO GET A 95% CONFIDENCE INTERVAL:

T-DISTRIBUTIONS LOOK ALMOST NORMAL, BUT THEY'RE FATTER OUT HERE.

$$\bar{x} \pm 2.26\left(\frac{s}{\sqrt{n}}\right)$$

WE USE A DIFFERENT **CUTOFF** DEPENDING ON WHICH T-DISTRIBUTION WE'RE USING AND HOW MUCH CONFIDENCE WE WANT.

FROM P. 201

INFERENCE ON A STANDARD DEVIATION

IN LITTLE SUZIE'S STORY, THE **CLT** CAN'T COME TO OUR RESCUE. THE SAMPLING DISTRIBUTION OF A STANDARD DEVIATION IS **NOT** GUARANTEED TO BE NORMAL, AND WE **CAN'T** COMPUTE A CONFIDENCE INTERVAL USING ANYTHING LIKE THE FORMULA ON PAGE 220 ABOVE. HOWEVER, **THE BASIC STEPS ARE THE SAME:** WE GENERATE A PARTICULAR KIND OF SAMPLING DISTRIBUTION (WITH SOME SNAZZY MATH) AND WE GENERATE A 95% CONFIDENCE INTERVAL BY CALCULATING THE PROBABILITIES UNDERNEATH IT (WITH SOME OTHER SNAZZY MATH).

THE SNAZZY MATH IS TOO COMPLEX TO DESCRIBE HERE, BUT IT'S ALL GENERATED USING THESE FUSE TIMES FROM 15 RANDOM DINGALINGS: 2.05, 2.25, 2.33, 2.40, 1.66, 2.39, 1.89, 2.18, 2.18, 2.06, 2.18, 1.89, 2.14, 2.38, 2.07. FROM THESE NUMBERS WE GET $\bar{x} = 2.14$, $s = 0.21$, AND $n = 15$.

NOTE THAT TO DO THIS KIND OF INFERENCE WE **HAVE TO ASSUME THE POPULATION IS NORMAL.** IN THIS CASE, IT MAY BE A FAIR ASSUMPTION, BUT IF THERE'S SOME KIND OF MANUFACTURING DEFECT THAT'S SKEWING THE OVERALL POPULATION (FOR EXAMPLE, AN UNEXPECTEDLY LARGE NUMBER OF FUSES FROM ONE FACTORY MIGHT BE DUDS, AND THIS ANALYSIS WOULDN'T ACCOUNT FOR THAT), OUR RESULTS **MIGHT BE MISLEADING.**

ALSO, LIKE THE T-DISTRIBUTION, THE SAMPLING DISTRIBUTION OF THE STANDARD DEVIATION CHANGES DEPENDING ON THE SAMPLE SIZE. AS IN ALL CASES, A LARGER n **WILL GIVE US MORE CONFIDENCE IN OUR CONCLUSIONS.**

FROM P. 202

CORRELATION

CORRELATION BEDEVILS LOTS OF STATISTICAL ANALYSIS. IF OUR SAMPLE MEASUREMENTS $x_1, x_2, x_3 \ldots x_n$ "CO-RELATE," THEY AREN'T INDEPENDENT, AND WE CAN'T USE ANY OF OUR STATS TOOLS ON THEM. SO WHENEVER DATA ARE CORRELATED, WE HAVE TO ACCOUNT FOR THE CORRELATION WHEN DOING INFERENCE.

FOR EXAMPLE, HEALTH STUDIES THAT INVOLVE **BIOLOGICAL TWINS** HAVE TO ACCOUNT FOR THE FACT THAT EACH TWIN'S DATA IS CORRELATED WITH THEIR SIBLING'S DATA, BUT WITH NO ONE ELSE'S. THIS CORRELATION CAN BE ELIMINATED USING WHAT'S CALLED A **PAIRED TEST.** OTHER CASES INVOLVE MEASUREMENTS THAT ARE CORRELATED **ACROSS GEOGRAPHICAL SPACE** (AS IN THE GECKO EXAMPLE ON PAGE 202), OR **ACROSS TIME** (IMAGINE A CASE WHERE AN INDIVIDUAL ALIEN'S SALIVA pH VARIES DEPENDING ON THE TIME OF DAY AND WE OBTAIN MULTIPLE MEASUREMENTS OVER TIME FOR THE SAME ALIEN). WE CAN OFTEN INCORPORATE THIS TYPE OF CORRELATION BY ANTICIPATING IT IN LARGER MATHEMATICAL MODELS. IN GENERAL, EACH TYPE OF CORRELATION REQUIRES ITS OWN SEPARATE TRICK.

FROM P. 203

REGRESSION ANALYSIS

IN GENERAL, WE USE **REGRESSION ANALYSIS** WHEN WE WANT TO EXPLORE THE RELATIONSHIP BETWEEN A **RESPONSE** VARIABLE AND ONE OR MORE **PREDICTOR** VARIABLES. IN THE EXAMPLE ON PAGE 203, SHRINKING MEDICINE DOSAGE IS THE PREDICTOR VARIABLE AND SHRINKAGE AMOUNT IS THE RESPONSE VARIABLE.

AFTER WE FIND THE **BEST-FITTING LINE** IN OUR SAMPLE DATA...

... WE USE IT TO BUILD A SAMPLING DISTRIBUTION THAT GIVES US A STATEMENT ABOUT THE POPULATION.

FOR EXAMPLE, IN THIS EXAMPLE WE'RE 95% CONFIDENT THAT A ONE-GRAM INCREASE IN DOSAGE...

... CORRESPONDS TO BETWEEN 2.3 AND 2.5 ADDITIONAL INCHES OF SHRINKAGE.

Shrinkage (inches) — 2, 4, 6, 8, 10, 12

Dose (grams) — 1, 2, 3, 4, 5

IN PRACTICE, WE HAVE TO BE CAREFUL WHEN WE'RE CURIOUS AS TO WHETHER PREDICTOR VARIABLES ACTUALLY **CAUSE** CHANGES IN RESPONSE VARIABLES. CAUSATION REQUIRES VERY CAREFUL EXPERIMENTAL DESIGN.

WE CAN ALSO USE THIS TYPE OF ANALYSIS TO COMPARE QUALITIES LIKE THE **HEIGHT** AND **WEIGHT** IN A POPULATION. IN THAT CASE, THE ESTIMATED SLOPE OF THE LINE WOULD TELL US ABOUT THE AVERAGE CHANGE IN WEIGHT FOR EVERY ONE-INCH INCREASE IN HEIGHT.

FROM P. 203

ANOVA

ANALYSIS OF VARIANCE (ANOVA) IS A HYPOTHESIS TESTING TECHNIQUE, BUT IT'S VERY DIFFERENT FROM THE HYPOTHESIS TESTING WE'VE LEARNED IN THIS BOOK. ANOVA WORKS BY COMPARING THE **VARIABILITY BETWEEN GROUPS** TO THAT **WITHIN GROUPS**. WE CAN USE IT WHEN WE WANT TO COMPARE **MORE THAN TWO** SAMPLE AVERAGES. THERE ARE **LOTS** OF WAYS TO USE ANOVA, AND HERE'S A QUICK RUNDOWN OF ONE:

I WANNA KNOW, DO THESE 5 SEPARATE FIRECRACKER SAMPLES ALL COME FROM A POPULATION WITH THE SAME AVERAGE...

...MAYBE FROM THE SAME FACTORY?

THE BOXPLOTS DON'T OVERLAP MUCH, SO I THINK NOT, BUT ANOVA CAN GIVE US A PRECISE ANSWER.

ANOVA FOCUSES ON THIS GENERAL QUESTION:

IS THE VARIATION **BETWEEN** THESE 5 SAMPLES...

...GREATER THAN THE VARIATION **WITHIN** THE SAMPLES THEMSELVES?

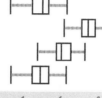

Blast Time (sec) 2 4 6 8 10

FROM P. 203

INFERENCE ON A PROPORTION

IF WE'RE CURIOUS ABOUT **WHAT PROPORTION** OF FOOTBALL FANS PREFER CHEESE DOODLES OVER PORK RINDS, OR **WHAT PERCENTAGE** OF LIKELY VOTERS PLAN TO CHOOSE TO RE-ELECT SENATOR SAM WARM IN THE UPCOMING ELECTION, WE CAN OFTEN USE THE BASIC INFERENCE STEPS WE'VE LEARNED IN THIS BOOK, BUT WE HAVE TO TWEAK THE DETAILS.

FOR EXAMPLE, HERE'S HOW WE COMPUTE A 95% CONFIDENCE INTERVAL FOR A POPULATION PROPORTION **p**.

WE USE OUR SAMPLE PROPORTION **p̂** (P-HAT) TO ESTIMATE THE POPULATION PROPORTION.

THIS IS THE STANDARD DEVIATION OF A SAMPLE PROPORTION.

WITH A LARGE SAMPLE SIZE, WE CAN USE A STANDARD NORMAL SAMPLING DISTRIBUTION. SO TO GET A 95% PROBABILITY, WE USE A CUTOFF OF 2 HERE.

$$\hat{p} \pm 2 \left(\sqrt{\frac{\hat{p}(1-\hat{p})}{n}} \right)$$

ONCE AGAIN, WE NEED A LARGE SAMPLE SIZE TO MAKE IT WORK, AND THE LARGER THE SAMPLE SIZE, THE NARROWER OUR INTERVAL!

IN OPINION POLLING WE REFER TO THIS AS THE **MARGIN OF ERROR**.

FROM P. 203

PREDICTING THE FUTURE

IN THIS BOOK WE'VE FOCUSED ON USING OUR SAMPLE DATA TO CHARACTERIZE ASPECTS OF AN OVERALL POPULATION, ITS AVERAGE, SAY, OR ITS STANDARD DEVIATION. HOWEVER, WE CAN ALSO USE STATISTICAL INFERENCE TO MAKE PREDICTIONS ABOUT **SINGLE OBSERVATIONS**. FOR EXAMPLE, WE CAN ASK QUESTIONS LIKE "BASED ON THE MEASUREMENTS I'VE GOT, WHAT'S THE **NEXT MEASUREMENT** (X_{n+1}) PROBABLY GOING TO BE?"

ONCE AGAIN WE CAN MAKE SOME HEADWAY USING OUR BASIC INFERENCE STEPS. FOR EXAMPLE, IF WE ASSUME THE OVERALL POPULATION IS NORMAL (ALWAYS A DICEY ASSUMPTION), WE CAN CREATE A "**PREDICTION INTERVAL**" THAT'S SIMILAR TO A STANDARD CONFIDENCE INTERVAL BUT A LITTLE BIT WIDER.

IN PRACTICE, STATISTICIANS DO THIS SORT OF THING TO PREDICT WEATHER PATTERNS OR FUTURE PRICES IN FINANCIAL MARKETS, THOUGH THEY USE MORE COMPLEX TOOLS.

A NOTE ABOUT THE AUTHORS

GRADY KLEIN IS A CARTOONIST, ILLUSTRATOR, AND ANIMATOR. HE IS THE COAUTHOR OF *THE CARTOON INTRODUCTION TO ECONOMICS*, VOLUMES ONE AND TWO, AND THE CREATOR OF THE *LOST COLONY* SERIES OF GRAPHIC NOVELS. HE LIVES IN PRINCETON, NEW JERSEY, WITH HIS WIFE AND TWO CHILDREN.

ALAN DABNEY, PH. D. , IS AN AWARD-WINNING ASSOCIATE PROFESSOR OF STATISTICS AT TEXAS A&M UNIVERSITY. HE LIVES IN COLLEGE STATION, TEXAS, WITH HIS WIFE AND THREE CHILDREN.